Microcrystal Polymer Science

Microcrystal Polymer Science

O. A. BATTISTA, Sc.D.

Vice President, Science and Technology, Special Consultant, Avicon, Inc.
Chairman of the Board and President, Research Services Corporation

McGRAW-HILL BOOK COMPANY

New York   St. Louis   San Francisco   Düsseldorf   Johannesburg
Kuala Lumpur   London   Mexico   Montreal   New Delhi
Panama   Paris   São Paulo   Singapore   Sydney
Tokyo   Toronto   Auckland

Library of Congress Cataloging in Publication Data

Battista, Orlando A
  Microcrystal polymer science.

  Bibliography: p.
  1. Microcrystalline polymers.   I. Title.
TP1183.M5B37   668.4'2   74-13742
ISBN 0-07-004084-2

1 2 3 4 5 6 7 8 9 0   KPKP   7 9 8 7 6 5

*The editors for this book were Jeremy Robinson, Stanley E. Redka,
and Lester Strong, the designer was Naomi Auerbach, and its production
was supervised by George E. Oechsner.   It was set in Baskerville
by York Graphic Services, Inc.*

It was printed and bound by The Kingsport Press.

*To seven associates whose vital support, at critical decision points
over a period of 18 years, made this book, the science and products
it introduces, and their promise for the future
the realities they are—at last!*

William C. Conner, LL.D.
Mamerto M. Cruz, Jr., Ph.D.
Herschel H. Cudd, Ph.D.
James M. Hait, D. Eng.
Merritt R. Hait, M.D.
Sherman K. Reed, Ph.D.
Frank H. Reichel, Jr., Ph.D.

Contents

Preface   *ix*

chapter 1 *Introduction* *1*

Part 1 *Natural Polymers*

chapter 2 *Microcrystalline Celluloses* *17*

      A. INTRODUCTION   17
      B. PREPARATION   21
      C. PROPERTIES   23
      D. APPLICATIONS   45

chapter 3 *Microcrystalline Collagens* *58*

      A. INTRODUCTION   58
      B. PREPARATION   61
      C. PROPERTIES   69
      D. APPLICATIONS   90

chapter 4 *Microcrystalline Silicates* *118*

      A. INTRODUCTION   118
      B. PREPARATION   120
      C. PROPERTIES   124
      D. APPLICATIONS   133

chapter 5   Microcrystalline Amyloses *138*

   A. INTRODUCTION 138
   B. PREPARATION 139
   C. PROPERTIES AND APPLICATIONS 141

Part 2   Man-Made Polymers

chapter 6   Microcrystalline Polyamides *149*

   A. INTRODUCTION 149
   B. PREPARATION 150
   C. PROPERTIES 152
   D. APPLICATIONS 164

chapter 7   Microcrystalline Polyesters *169*

   A. INTRODUCTION 169
   B. PREPARATION 170
   C. PROPERTIES 175
   D. APPLICATIONS 180

chapter 8   Microcrystalline Polyolefins *182*

   A. INTRODUCTION 182
   B. PREPARATION 183
   C. PROPERTIES 184
   D. APPLICATIONS 190

chapter 9   Literature Cited *193*

   appendix   197

   index   201

The purpose of this book is to introduce a new field of polymer science, one which transcends and joins together two major disciplines—conventional high-polymer science and colloidal science. It is intended to bring together original research on microcrystal particles recovered from the major classes of linear high polymers, both natural and man-made.

Microcrystal polymer science as defined and delineated in this book became a commercial reality with the introduction of microcrystalline celluloses (Avicels) in 1961, colloidal (nonfibrous) forms of cellulose now sold throughout the world and manufactured in plants in the United States and Japan.

Polymer microcrystals as described within the context of this book are discrete particles extracted from within the matrix of polymers, particles that have maximum dimensions varying from 50 to less than 10,000 Å, which is the size range for most viruses and many smaller bacteria.

Research on microcrystal polymer products then advanced to include collagen, mineral silicate (chrysotile asbestos), amylose, polyamides, polyesters, and polypropylenes. The results of this extensive research are reported in individual chapters, each concerned with the respective major conventional polymer composition in the order listed above.

The preparation, properties, and uses of each microcrystal polymer species are covered in each product chapter, along with extensive electron micrographs (many published for the first time) where applicable.

The text is addressed to research scientists and engineers who are involved in the research, development, and manufacture of high polymers and to teachers and students interested in the disciplines of both polymer and colloid chemistry. It has been organized in a manner which should permit its adoption as an introductory course in a new branch of science with the existing curricula on polymer chemistry or colloid chemistry.

A special effort has been made to make the book as readable as possible with effective illustrative supports so that it may have value and interest for those involved in the peripheral facets of polymer and colloid science.

For example, microcrystal polymer science offers a new frontier of investigation of interest to the biological, biochemical, and medical disciplines and introduces a new class of organic crystals whose potential for the physicists still is at a level of theoretical interest.

A treatise such as this which covers a period of about 20 years of research necessarily evolves out of the contributions of many people. In addition to the gentlemen to whom this book is dedicated and to my present employer, Avicon, Inc. (owned 50/50 by FMC Corporation and Alcon Laboratories, Inc.), I wish to express my appreciation for the assistance of so many who have in one way or another assisted me in the writing of this book or in the procurement of the data within its covers.

I am especially indebted to the following persons for their support and friendship as associates or as coworkers and for the privilege in a few instances of including some results of their laboratory work in this treatise: Mr. Fred M. Bowers, Dr. Ernest F. Couch, Dr. N. Z. Erdi, Mr. Frank J. Karasinski, Mr. George F. Leone, Dr. Egon Matijevic, Mr. Frederick F. Morehead, Dr. George F. Mueller, Dr. N. N. T. Samaras, Mr. William G. Strunk, Dr. Charles S. Venable, Mr. Richard L. Ward, and Dr. William H. Watson.

To all three of my secretaries over the entire period covered in the creation of this work—Mrs. Dorothy D. Dunkin (1955 to 1962), Mrs. Berenice M. Feaster (1963 to 1970), and especially to my present secretary, Mrs. Diane A. Stegall, CPS, who since 1971 has been particularly competent and dedicated in typing the manuscript for the treatise and helping with some of the editing to put it into final form—my lasting thanks.

A special acknowledgment also is gratefully recorded to my wife, Helen Keffer Battista, and my children, William and Elizabeth Ann, who have always been so patient, cooperative, supportive, and understanding of the inordinate percentage of time that I have devoted to my research and writing over a span of more than 25 years.

O. A. Battista

CHAPTER ONE

Introduction

For many years a large number of investigators have studied the dimensions, the surface charges, and the morphology or architectural order of microcrystals extracted from cellulose. Electron microscopy, x-ray diffraction, and a wide variety of kinetic data provided valuable insight into the fine structure of the precursor celluloses and their physical properties.

Microcrystal polymer science, the primary subject of this treatise, might be defined as *systems of colloidal-size polymer microcrystals whose suspensoid properties are largely determined by the relative proportion of discrete unit particles in suspension versus the proportion of the same particles present as aggregates, each aggregate containing varying numbers of the same microcrystals clustered together.*

Colloidal microcrystal polymer science (*macrocolloidal science*) embodies crucial differences from the phenomena of conventional colloid science. The word *colloid*, coined as a branch of science by Thomas Graham in 1861, comes from the Greek word *kolla* meaning glue. In the latter instance, properties are largely determined by the fact that a significant proportion of molecules lie in or are associated with interfacial regions. It was long recognized such as with gold particles in suspension that subdividing a bulk element into ultrafine particles changed its properties in interesting ways.

Macrocolloidal size as herein advanced embodies these premises:

1. Each unit particle (microcrystal) consists of a cluster of many long-chain molecules held together laterally by noncovalent bonds (a simple definition of a polymer microcrystal) in which the polymer molecules are crystallized in the folded state, the extended linear (fibrous) state, or gradations thereof depending on the "genetics" of the polymer precursor.

2. Free molecules or residual homopolymer contaminants are substantially absent because the microcrystal particles are selectively extracted from a predetermined and preexisting polymer matrix.

3. The properties of microcrystal-gel suspensoids are related to the proportion of unit microcrystals and aggregated microcrystals present.

As the course of the research with polymer microcrystals progressed, it became increasingly evident that the surface interactions between discrete microcrystals of varying sizes, their degrees of perfection, their size distributions, and their topochemical functional groups all entered into the unique properties of the many families of suspensoids studied. In more recent years, property differences became delineated depending on whether the molecules in the microcrystal existed in the folded form or in the extended fibrous form.

Staudinger first observed single crystals of organic polymers in his work in the polymerization of formaldehyde (1).* A little later Schlesinger and Leeper (2), working with gutta-percha, also observed areas of crystallinity or high lateral order. The advent of linear polyethylene led to a more rigorous study of single polymer crystals and opened up an era of intensive scientific study on such products. This activity reached its peak of intensity during the decades 1950 to 1970.

The accidental discovery of stable, opaque gels of pure-cellulose microcrystals (3, 4) at high-solids concentrations, upon which a new industry has evolved, occurred in 1955. The writer, while at American Viscose Division of FMC Corporation, was attempting to isolate tiny crystals of cellulose in quantity with a plan for using them to nucleate the viscose solution from which rayon tire yarn is produced. It was projected that this would permit greater control of the cellulose structure during the spinning process, which enables the manufacture of a better tire yarn for use as tire cord.

The diverse commercial promise of the microcrystalline cellulose gels at high solids stemmed from their unusual viscosity properties and, initially, from their potential use as edible and completely calorie-free food ingredients. These suspensoids spread like butter but would not melt; they became fluid when shaken but recovered thickness quickly when allowed to stand. Their thixotropic and pseudoplastic thickening properties were most unusual. These and other considerations led to the initiation of the expansion of the research on cellulose to other crystalline linear polymer precursors (9), which has continued with increasing emphasis since 1955. Out of this effort there has emerged a unifying concept that has found experimental verification when applied to several crystalline and linear high-molecular-weight polymer precursors in addition to cellulose.

The list of insoluble microcrystal polymer products from both natural and

*For references, indicated by numbers in parentheses, see the section Literature Cited at the back of the book.

man-made linear polymer precursors that now have been converted into the microcrystalline colloidal state includes in addition to cellulose, amylose, magnesium silicate (asbestos), collagen, natural silk, wool, polyamides, polyesters, polycarbonates, and isotactic polypropylenes.

If you use Fig. 1.1 as an outline, it is reassuring to examine the more significant commercial, demonstrated, and/or technically feasible uses projected at this time for some of these new polymer compositions, polymer compositions whose competitive utility depends upon the inherent properties which their constituent, discrete, colloidal microcrystal polymer particles (crystals) give them.

The author makes these projections, albeit with some measure of brazenness in anticipating some of the yet-to-be-realized commercial developments, because he believes that only a small crack in the door of this evolving new field of macrocolloidal science has been made.   What better reason may one have to pursue this beginning to a more mature, more universally rewarding state of knowledge?

1. Microcrystalline celluloses—major uses
    a. Tablet binder and excipient and suspending agent for pharmaceuticals
    b. Control of "heat shock" in frozen desserts

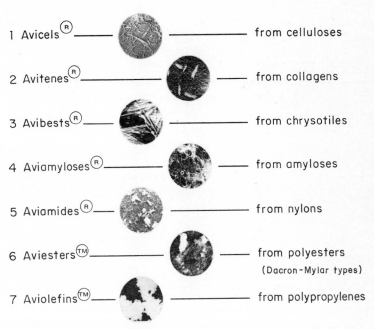

1 Avicels® ——————————— from celluloses

2 Avitenes® ——————————— from collagens

3 Avibests® ——————————— from chrysotiles

4 Aviamyloses® ——————————— from amyloses

5 Aviamides® ——————————— from nylons

6 Aviesters™ ——————————— from polyesters
                                    (Dacron-Mylar types)

7 Aviolefins™ ——————————— from polypropylenes

fig. 1.1   *Microcrystal polymer products.*

  c. Canned convenience foods (meat, fish, poultry) containing stable heat-sterilized salad dressings

  d. Rheology-control agents in paints, ceramics, and the manufacture of decorative laminates

  e. Special grades that act as emulsifiers of many products without changing surface tension

  f. Effective substrate for thin-layer chromatography and column chromatography

  g. Precursor of unique structural products from which spheres or massive carbon and/or graphite structures may be produced

2. Microcrystalline collagens—major uses (See Table 3.16 for more extensive list)

  a. Bioassimilable hemostat adhesive

  b. Wound and burn dressings

  c. Bioassimilable bone and prostheses

  d. Sutures

  e. Capsules

  f. Dialysis membranes

  g. Photography

  h. Cosmetics

3. Microcrystalline silicates—major uses

  a. Industrial rheology control and suspending agent

  b. Catalyst support and binder

  c. Reinforcement fibrils for plastics and soft metals that melt below 660°C

  d. Fireproof 100 percent inorganic papers

4. Microcrystalline starches—major uses

  a. Food uses similar to those for microcrystalline celluloses except that microcrystalline amyloses have normal caloric content for starches of 4 cal/g

  b. Some medical uses similar to microcrystalline collagens in view of their bioassimilability

5. Microcrystalline nylons—major uses

  a. Aqueous suspensoids for coatings and related applications, such as coatings for glass fibers, aluminum, etc.

  b. Ultrafine colloidal particles for fluidized-bed dry-powder coating applications

  c. Rheology control and pigment dispersion aids for paints and inks

  d. Fused films with controlled ultrafine pore sizes for ultrafiltration, membranes, etc.

  e. Production of ultrathin self-supporting nylon films

6. Microcrystalline polyesters—major uses

  a. Viscosity stable, highly thixotropic, aqueous suspensions for coatings and related applications where an acid-and-alkali-resistant film is needed

  b. Ultrafine colloidal powders for fluidized-bed dry-powder coating applications
7. Microcrystalline polyolefins—major uses
    The projected uses for these products still in the developmental stage are similar to those for microcrystalline nylons and microcrystalline polyesters, respectively, except that the primary microcrystalline polyolefin product is derived from isotactic grades of polypropylene.

The initial experiments from which microcrystalline cellulose was produced used rayon cellulose as the raw material. Rayon tire cord was converted to its level-off degree of polymerization (LODP)* (3) by means of strong hydrochloric acid (2.5 N HCl at 105°C for at least 15 min).

In attempts to produce working quantities of these 75- to 100-Å cellulose microcrystals, LODP cellulose from rayon was subjected to severe mechanical disintegration in water at 10 percent solids for 30 min. Colloidal suspensoids that looked like opaque petrolatum were produced (Fig. 1.2). Such gels at low magnification reveal clearly birefringent large aggregates in suspension (Fig. 1.3). Electron microscopy of the supernatant is required, of course, to observe the individual cellulose microcrystals in such a suspensoid, and such gels are clearly composed of a highly polydisperse distribution of cellulose microcrystals and aggregates thereof.

Mechanical disintegration of pulverized-wood-pulp powder in water at high solids leads invariably only to a two-phase separation (Fig. 1.4), and the particles clearly remain as intact fiber fragments without the release of discrete unhinged microcrystals (Fig. 1.5).

*This term refers to the number of repeating monomer units calculated from the weight-average molecular weight of the macromolecules that make up the discrete, colloidal, unhinged microcrystals of polymers.

fig. 1.2 Microcrystalline cellulose (15 percent gel).

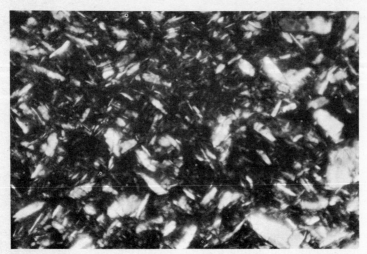

fig. 1.3 *Microcrystalline cellulose gel (14 percent solids) as a thin smear on microscope slide.*

On the other hand, when LODP cellulose (microcrystalline cellulose) is subjected to the same treatment in a Waring Blendor for the same lengths of time as pulverized α-cellulose wood pulp, increasingly stable and highly viscous suspensoids result (Fig. 1.6).

High-solids microcrystalline cellulose gels possess unique and useful properties in contrast with the dilute monodisperse transparent or semitransparent dilute suspensions observed by other investigators who had studied single cellulose microcrystals (6) or peptized species of cellulose crystals such as those

fig. 1.4 *Powdered cellulose after Waring Blendor treatment for (left to right) 1, 5, 10, 15, 20, and 30 min.*

fig. 1.5 α-Cellulose pulverized at 10 percent solids in a Waring Blendor for 30 min.

produced using strong sulfuric acid (7, 8). Practical uses of the dilute suspensions of cellulose microcrystals which many investigators, including ourselves, prepared for electron-micrographic crystal structure studies never were evident or projected for such dilute suspensions.

Of crucial significance with respect to the evolution of this treatise was the realization that the high-solids polydisperse cellulose microcrystal gels had commercial value, and aggressive steps were initiated to demonstrate their marketability (3, 9).

The field of colloidal microcrystal polymer science with which this treatise

fig. 1.6 Microcrystalline cellulose at 10 percent solids after Waring Blendor treatment for (left to right) 1, 5, 10, 15, 20, and 30 min.

is concerned derives its originality from the following combination of pre-requisites.

1. The starting material for each product must be a high-molecular-weight polymer, a compound composed of linear long-chain molecules. These long molecules, when allowed by nature (natural polymers) or by man (synthetic polymers) to crystallize as synthesized *in situ*, precipitated from solution, or crystallized by cooling a melt (in the case of thermoplastic long-chain polymers) are the bases upon which all fibers, films, and many structural plastics are produced. The manner in which the long-chain molecules crystallize will depend on their environment at the time of aggregation: e.g., whether they will line up side by side like matchsticks to form *fibrous* microcrystals or whether the molecules fold on themselves and pack into lamellar-like microcrystals.

In any case, the same molecule may pass from one microcrystalline area or physical phase to several others, depending on the polymer species, the original length of the molecules, the thermodynamics of the crystallization environment, etc. Preconditions must exist in the polymer precursor to permit some sites or regions of imperfection (chemical or physical) to become selectively attacked by a chemical or another degradative means. Naturally, the exact loosening mechanism varies with each polymer precursor, depending on its characteristic morphology and the chemical nature of its constituent macromolecules.

The prerequisite is to cut, etch, or otherwise weaken such sites without extensively destroying the integrity of the original ordered regions in the process. A minimum number of the microcrystals must then be separated from each other in water or in another low-swelling medium to produce the new colloidal polymer gels. If too few are freed, a "sand-in-water" separation is effected.

The unhinging or etching is usually done by a specific chemical attack which varies for each polymer system. Hydrochloric acid under controlled conditions is used to disconnect the microcrystals embedded in cellulose fibers, in amylose, and in polyamide matrices. Mechanical agitation from a device that provides a high rate of shear, such as a blender, is used to assist in the separation of microcrystals from their unhinged but clustered-together state.

Within the framework of the foregoing concept, a colloidal polymer microcrystal can be defined as any crystalline, highly ordered, or discrete colloidal particle separated from its position within the matrix of a polymer-precursor product. These particles possess a maximum dimension (truly colloidal) less than 10,000 Å, more commonly less than 5,000 Å, and minimum dimensions as small as 50 Å. Furthermore, the size of the ultimate microcrystal or colloidal particles corresponds closely to the dimension that they possessed in the original polymer-precursor matrix prior to the chemical pretreatment and subsequent deaggregation process.

2. The major innovation is that the production of colloidal microcrystal polymer suspensoids as have been developed represents the reverse of the conventional way of producing individual crystals, specifically by aggregating molecules from solution or from molecular dispersion in the molten state through coagulation, precipitation, and crystallization. Such polymer microcrystal systems may, of course, be derived from synthetic and/or natural organic as well as inorganic polymer precursors.

3. The microcrystals to be unhinged or loosened and subsequently dispersed are bonded by interconnecting molecular chains (covalent bonds) within the matrix of the polymer precursor.

4. To achieve the unique colloidal microcrystal gels it is necessary to liberate a sufficient number of single crystallites, or regions of highest lateral order, from within the matrix that forms the original polymer precursor. Once the appropriately controlled or selective degradation or etching of the interconnecting hinge areas has been accomplished, under conditions that preclude excessive swelling of the constituent crystallites or regions of highest lateral order, the microcrystals are released into a liquid medium by mechanical energy. Of particular importance in this regard is that this route to the isolation of polymer microcrystals substantially precludes the occlusion of small amounts of low-molecular-weight homopolymer or other by-product contaminants that are bound to be present when microcrystals are produced by conventional polymerization techniques such as emulsion polymerization of monomers, etc. The advantages of this route in precluding harmful effects on properties due to very low-molecular-weight residues or contaminants have been borne out in practice.

5. The presence of a minimal concentration of single colloidal microcrystals in association with much larger colloidal particles comprising aggregations of the aforementioned basic unit microcrystal is a prerequisite for the formation of stable polymer microcrystal gel systems possessing unique rheological properties. The straightforward disintegration of a fibrous polymer precursor in which the microcrystals have not been previously disconnected or *loosened* will not result in the formation of similar stable gels possessing valuable high initial yield values and extreme viscosity shear-stress dependencies. Furthermore, a sufficiently high concentration of solids must be present during the mechanical disintegration step (usually at least about 5 percent) to obtain gels that exhibit the unique functional and rheological properties characteristic of these suspensoids. In this respect, microcrystal polymer suspensoids differ from conventional gels in which the network of chain molecules is held together either by primary bonds or by thermally reversible linkages that are due to secondary attractive forces.

The rheological properties of the microcrystal gels depend mostly on the particle size distribution, particularly on the fraction under 1 μm in maximum dimension. The microcrystal gels are characterized by unusual high yield values resulting from a three-dimensional brush-heap (insoluble microcrystals touching each other) network. This structure breaks down instantaneously as the elastic limits of the network are surpassed and will rebuild at velocities that are different for each microcrystal gel. The apparent viscosities, of course, decrease sharply with increasing rates of shear.

The microcrystals pervade the whole system at low concentrations without settling. Their three-dimensional network has a certain rigidity, and, consequently, they should be characterized as gels. Beyond the yield value, increasing rates of shear lead to decreasing apparent viscosities. The gels liquefy reversibly upon shaking and solidify upon standing, the process being time-dependent. Thus all microcrystal gels appear to be thixotropic and pseudoplastic with a yield point. Moreover, rheological studies indicate that they have an irregular, time-dependent, extreme yield point, and portions of their flow curves show rarely encountered pseudoplastic-rheopectic behavior.

One common feature of these gels is that the microcrystals as finally recovered from each polymer have on their surface polar groups such as —NH$_2$, —OH, or —CONH— which are capable of hydrogen bonding. Hence particle-particle interaction or particle–solvent-particle type of bonding has to be assumed as a significant operating force of gelation. It is likely that the small spheres and ellipsoids first form chains of beads which branch off and subsequently pervade the system, whereas the long, rod-shaped or fibrous microcrystals are probably arranged in the gels like architectural scaffolding or heaped matchsticks. Such arrangements assure the stability of the three-dimensional network.

Our electron microscopy and x-ray diffraction studies have revealed that the size and shape of the smallest recoverable microcrystalline particles from the linear polymer precursors are remarkably similar in size and shape for several widely different polymer species. Furthermore, recovery of the *smallest* microcrystal unit component is highly dependent on the severity of the mechanical disintegration treatment. In other words, macromicrocrystals recovered under relatively severe conditions of mechanical shear may appear as substantially linear or *fibrous* particles or as *flat platelets,* whereas by subjecting them to extremely severe conditions of mechanical shear at high solids, it seems possible to reduce the macromicrocrystal entities into smaller subunit particles having diameters in only the 100- to 200-Å size range. Several electron micrographs revealing these findings are shown in this treatise, and work is continuing with respect to the significance of these observations vis-à-vis conventional understandings of the ultrafine structure of polymer microcrystals.

Flat, lamellar-type polymer microcrystals usually develop when the molecules crystallize in the folded-chain configuration from very dilute solution. Under these conditions, the long-chain molecules may crystallize into extremely thin wafers of folded chains only about 100 to 150 Å thick. Such tiny platelets possess a tremendous surface area; a relatively simple calculation reveals that the surface area of a single gram of 100-Å-thick platelet polymer crystals might possess a surface area as high as 1 million cm^2.

It is well recognized, of course, that every polymer microcrystal must have originated from a nucleus. The tiniest nucleus would be an infinitesimal crystal and have almost no morphology but essentially all surface. On the other hand, most polymer microcrystals are giant crystals vis-à-vis crystals in the conventional crystallographic sense. Inasmuch as this treatise is being restricted to polymer crystals derived from high-molecular-weight linear polymer precursors, a quotation by Dr. Bernard Wunderlich (10) is especially appropriate, as it clearly illustrates the tremendous importance of specific surface area of particles on physical properties of mixtures.

> Consider a typical linear polymer molecule magnified 10 million times. The length of one molecule is 100 feet and the diameter becomes 0.125 inch. About 20,000 to 30,000 backbone atoms are already interconnected in such a molecule. The difference in bond energy along the chain (approximately 83 kcal. per mole) is so much larger than that of bonds between the chains (approximately 0.3 kcal. per mole) that, at the crystallization temperature, there is no chance of breaking any of the backbone bonds. The difference between crystallizing methane and polyethylene is similar to the difference in handling peas and 100-foot long cooked spaghetti.

Relatively free-moving linear polymer molecules, when they are precipitated from solution or congealed from a melt, sometimes, depending on the environment, crystallize in a folding pattern rather than side by side like wooden matchsticks. In other words, the chains may be folded back and forth every 100 Å or so instead of aggregating laterally like long strands of licorice stuck together in bundles. The folding usually takes place because it may be faster for the long flexible single-chain molecules to crystallize in this manner, even though extended linear long-chain molecules are thermodynamically more stable than folded-chain molecules. Folding requires energy. Of course, once chains become folded within the morphology of a crystal, it becomes most difficult if not impossible to introduce enough energy to realign the folded molecules into the more thermodynamically stable linear (fibrous) crystal morphology.

It is now understood why most linear polymers, especially those of high molecular weight, never achieve 100 percent crystallinity. And this is a good thing! If they did develop *perfect* crystallinity, their resulting physical properties in many cases would cause them to perform in quite a different manner from the widely useful end products they presently are. As a matter of fact, one of the reasons that polymer microcrystals in general are relatively small is because it becomes increasingly difficult to achieve more perfect alignment of long-chain molecules as the final size and degree of order of the crystal increases.

In summary, the science treated in this book is not directly concerned with the conventional route of synthesizing long-chain molecules, which is to put them into solution, melt them, or otherwise reconstitute them into fibers, films, or other structural forms. Rather, this science starts with a structure, schematically not unlike a brick wall in some respects, and etches away or loosens the mortar between the bricks under conditions that do not dissolve or swell the bricks. The main difference, of course, is that the bricks that are freed are so tiny that they are invisible to the human eye, and the mortar removed or loosened consists of fragments of molecules that intertwine in some way between the bricks.

Furthermore, the final size of the bricks is predetermined by a combination of the history of the polymer precursor, by the conditions of attacking the matrix in which they are embedded, and especially by the intensity and severity of the mechanical shear energy used to separate them. It has been found that these little bricks or microcrystals can vary in size from that of a rabbit papilloma virus up to that of a tobacco mosaic virus. Interestingly, as with the viruses, they may be round or rodlike microcrystals. The shape of the smallest basic unit microcrystal recoverable depends on the molecular geometry within the precursor polymer and the severity of the mechanical treatment used to dislodge the unhinged microcrystals from within their preestablished matrices.

Polymer scientists have demonstrated an outstanding track record in developing products tailor-made to meet specific end-product uses with commercial price and performance advantages. And out of the intense competitive scramble there has emerged just a handful of predominant polymer species that have stood the tests of time in meeting the majority of man's needs. Among these fibers the celluloses, nylons, polyesters, polypropylenes, and the polyethylenes stand out. Among the plastics and structural products the polyethylenes, the polyvinyl chlorides, and polystyrenes now play a leading role in the major countries that produce polymers. The polyolefins—polyethylene and polypropylene—are out front as the most widely produced plastic materials in the world.

Striking evidence of the outgrowth of all the research and development on polymers that has ensued since the 1930s is reflected in a single statistic. Currently, almost 40 billion lb of man-made fibers, films, plastics, and elastomers—all polymer products—are manufactured each year in the United States alone. Conventional polymer science and technology during the past 40 years has produced an almost endless variety of useful materials. Market researchers are predicting that output of the polymer industry will exceed that of all the metal industries by 1985.

By comparison, colloidal (rather than fiber, film, plastic, and/or elastomer products) polymer products as defined in this treatise are relative newcomers to the commercial scene. The first of these commercial products—microcrystalline cellulose—has reached annual sales of many millions of pounds.

Microcrystalline collagens have been tailored to function in surgery as new and more effective bioassimilable hemostatic agents. Their many potential uses are only at the "toe" of the exponential curve of commercial development. Other microcrystal polymer products—microcrystalline amyloses, microcrystalline polyamides, microcrystalline polyesters, and microcrystalline polypropylenes, for example—remain farther back in the wings awaiting their respective turns for future utilization. That their turn will come there is little doubt; the only things holding up their commercial development for a wide range of medical, cosmetic, and industrial uses are the time and money that must be earmarked and expended to prove their advantages in the marketplace. This author believes firmly that the growing commercial successes of the initial offspring of this family of new polymer products will provide the momentum that will permit the commercial fruition of the more recent arrivals within the family of microcrystal polymer products.

PART ONE

Natural Polymers

CHAPTER TWO

Microcrystalline Celluloses

INTRODUCTION

From the time that the term *hydrocellulose* was coined in 1875 (11) up until the introduction of commercial colloidal forms of microcrystalline celluloses in 1962 (3), the production of useful products by the degradation of celluloses by acids seemed like the wrong direction to go. By going in this direction with unrelenting persistence, however, entirely new commercial uses for cellulose, nature's most abundant plant polymer, have become a reality.

Historical uses of cellulose have been dependent largely upon its fibrous nature, either in its natural or regenerated state. Microcrystalline cellulose—and the growing family of other microcrystalline polymer products that have emerged as its progeny—are colloids with all of the proven and implied possibilities described in this treatise.

The first references to the existence of definite crystalline zones interposed in the amorphous structure of cellulose materials were made by Nageli (12), who in 1877 already confirmed the optical anisotropism of vegetable products both in cell walls and in fibers. Herzog (13) and Meyer (14) postulated a crystalline morphology for pure cellulose. The structure of the elemental crystallite in native cellulose was formulated by Meyer and Misch (15), who considered it to be made up of five units of cellulose molecules or cellobiose radicals arranged in parallel. The length of the crystallite along the *b* axis is 10.3 Å, precisely the length corresponding to the cellobiose unit (16) made up of two anhydroglucose residues, whereas that of the *a* axis is 8.35 Å, a sufficiently small distance so that the space between the proximal —OH groups of different chains offers the possibility of the formation of strong hydrogen bonds in accordance with Pauling's calculations (17) for this type of bond.   Later structural studies, following Meyer's proposals, by Hermans and de Booys (18) and Van der Wyk and Meyer (19) led to very similar conclusions for the dimensions of the unit cell for native cellulose—dimensions generally accepted today.

Crystallite zones also appear, of course, in regenerated celluloses, in mercerized celluloses, and in alkali celluloses.   Hess (20) and Meyer and Badenhuizen (21) proved that the crystallites of the above forms of cellulose differ from those for native celluloses in size as well as in the angles of the planes of the atoms within the molecules of their unit cells.

Commercial forms of microcrystalline celluloses as new, stable thixotropic-gel systems involving aqueous colloidal dispersions of disintegrated level-off DP cellulose, abbreviated as $\overline{DP}$ (3), at high-solids concentration were first described in a patent issued in 1961 (4).

This patent, along with a later publication (3), described a combination of two characteristic prerequisites for producing novel colloidal phenomena from a fibrous high polymer such as cellulose: (1) a controlled chemical pretreatment to destroy the molecular bonds whereby microcrystals are hinged together in a network structure, and (2) appropriate use of mechanical energy to disperse a sufficient amount of the unhinged microcrystals in the aqueous phase to produce the characteristic novel rheology and the smooth, fatlike spreadability of the resulting colloidal microcrystalline cellulose gels. It was clearly demonstrated that stable gel systems were obtained only when the mechanical energy was introduced into an aqueous suspension of level-off DP cellulose in which the total solids concentration was of the order of 5 percent or more and only if the mechanical energy was severe enough to liberate a minimum number of microcrystals to make a stable gel possible.

Although many researchers, including ourselves, had studied the isolation of microcrystals (level-off DP cellulose) from cellulose fibers, these studies were directed primarily at achieving a better understanding of cellulose fine structure.   For this reason most of the early investigations of cellulose microcrystals using electron microscopy and x-ray diffraction techniques involved colloidal dispersions of microcrystals of cellulose at very low concentrations. In this area special mention should be made of the work on colloidal cellulose microcrystals by Ranby (6), Morehead (22), Mukherjee and Woods (8), and Marchessault et al. (7).

For example, Mukherjee and Woods isolated microcrystals using strong sulfuric acid under conditions that peptized the microcrystal surfaces, no doubt owing to limited formation of sulfate-ester groups on the surface of the microcrystals.   Such highly peptized and surface-modified (esterified) microcrystals are extremely difficult to purify and, from a practical viewpoint, do not give the same visual and functional properties as underivatized cellulose microcrystals produced with hydrochloric acid. The hydrodynamic characteristics of colloidal dispersions of microcrystals produced by various severe acid-digestion treatments were studied by Marchessault et al. (7), by Hermans (23), and by Hermans and Edelson (24). Numerous other investigators have been concerned with the hydrolysis of cellulose, and their work has been reviewed previously (25).

Highly dilute dispersions of cellulose microcrystals did not yield products with functional and rheological properties of demonstratable commercial value. Nor did they respond to efficient deaggregation techniques for microcrystal separation unless partially derivatized as by reaction with sulfuric acid. The use of a nonpeptizing acid such as HCl in combination with a minimum concentration of solids during the mechanical disintegration step was found to be critical to achieve useful products by a practical commercial process (3, 4).

Good correlation between the lengths of the microcrystals as measured from electron micrographs of the individually dispersed particles and the weight-average molecular weights of the constituent molecules in the microcrystals was found over a wide range of microcrystal lengths (5). In addition, in the case of rodlike crystals recovered from such native celluloses as wood or cotton it was observed that ultrasonic treatment of aggregated particles released individual microcrystals without decreasing their length (22), which demonstrated that ultrasonic treatment caused only lateral deaggregation of the already unhinged microcrystals. The microcrystals, provided they are not topochemically derivatized, have been shown to be negatively charged (23, 24).

More recently, Matijevic et al. (26), working with cellulose microcrystals recovered from wood pulp, investigated the coagulation effects on microcrystalline cellulose sols of simple ions, hydrolyzed metal ions, chelate complexes, and a series of alkylammonium chlorides differing in length of the hydrocarbon chain. It was found that microcrystalline cellulose sols are much more sensitive toward electrolytes than typical lyophilic colloids. The coagulation concentrations of simple monovalent counterions were even considerably lower than for many lyophobic sols. It was suggested that this is due to the second stability minimum as predicted by the theory of Derjaguin-Landau-Verwey-Overbeek (DLVO). Reversal of charge of the microcrystalline cellulose particles could be achieved neither with complex counterions nor with the surface-active ions employed. All these are efficient reversal-of-charge agents when added to *typical* lyophobic colloids.

The size and shape of microcrystals in stable aqueous colloidal dispersions of cellulose are largely determined by the nature of the starting raw material, especially by whether the cellulose is in its native or regenerated state (5). For example, a representative spectrum of the size of such microcrystals is given in Table 2.1. The average length of each type of microcrystal corre-

**table 2.1   Level-off Basic Degree of
Polymerization of Natural and Regenerated Fibers (5)**

Form of cellulose	Average basic $\overline{DP}$ range
Natural fibers:	
Ramie, hemp...	350–300
Cotton, purified..	250–200
Unbleached sulfite wood pulps	400–250
Bleached sulfite pulps..	280–200
Bleached sulfate wood pulps	190–140
Mercerized cellulose (18% NaOH at 20°C, 2 h)...........	90–70
Vibratory-milled wood cellulose	100–80
Regenerated fibers:	
Fortisan...	60–40
Textile yarns..	50–30
Tire yarns...	30–15

sponds approximately with the measured level-off DP of each sample. The level-off DP of wood-cellulose microcrystals is 220, whereas the level-off DP of the cellulose in rayon-tire-cord microcrystals is only about 30. Stable microcrystalline cellulose gels and dispersions may be made from all types of level-off DP cellulose (3, 4).

The distribution of the size of microcrystal particles in microcrystalline cellulose gels is normally quite heterogeneous. Only by ultracentrifugation at low concentrations can a fraction of homogeneous size distribution be obtained (23). Commercial pure microcrystalline cellulose gels may have only 20 to 30 percent of individually dispersed microcrystals; the remainder are made up of aggregates of unhinged microcrystals as large as 10 to 50 μm (3). It is important therefore to recognize this fact in interpreting data on the measured properties of microcrystalline cellulose gels.

The surfaces of the cellulose microcrystals are saturated with hydroxyl groups, whereas the —CHO groups are concentrated more on the ends of the microcrystals, which reflects the hydrolytic unhinging of the 1,4-β-glyco-sidic bonds in cellulose molecules that originally connected the regions of high lateral order together (3). Structural forms of cellulose with unique thermal resistance and a marblelike physical appearance, produced by air-drying microcrystalline cellulose gels at atmospheric pressure, can be explained only by hydrogen-bonding forces engaging laterally as water is evaporated, with the assistance of powerful surface-tension forces (3, 4).

Jayme and Knolle (27) have provided evidence to suggest that some rodlike cellulose crystallites prefer to aggregate longitudinally. It is suggested that under some circumstances the ends of the crystallites are more susceptible to bond formation than the sides. On the other hand, Battista and Smith

(3) found that by spray-drying high-solids (9 percent) microcrystalline cellulose gels the originally unhinged microcrystals were made to aggregate into particles resembling a ball of twine in which the microcrystals were randomly arranged like a pile of wooden matchsticks with holes and crevices of 10 to 100 Å between the agglomerated particles.

PREPARATION

When fibrous cellulose fibers such as cotton or wood pulp are beaten in water (the key step of all papermaking processes with cellulose fibers), very long strands or fibrils of cellulose are formed. Within each long strand or fibril of cellulose lie numerous microcrystals interconnected by cellulose molecules. The microcrystals in the cellulose fibrils are therefore hinged together by very strong forces because the molecules that weave in and out of them are connected by covalent 1,4-β-glycosidic linkages.

Mechanical energy alone is not an efficient or effective means of isolating polymer microcrystals essentially as they are preformed in a polymer-precursor matrix. Pulverized fibrous cellulose which has not been unhinged with acid does not produce a stable dispersion. Cellulosic gums also do not resemble the new microcrystalline dispersions.

Initially, acid attacks cellulose rapidly. Mild hydrolysis breaks some of the covalent bonds, and recrystallized areas may appear. Severe acid hydrolysis breaks all the molecular hinges and reduces the fiber to disconnected microcrystals, provided the acid does not overpeptize, swell, or dissolve the original microcrystals.

The reaction rate levels off at this stage. The microcrystals are fairly equal in size for a given cellulose raw material, although size distribution of microcrystals in cotton is more uniform than in wood pulps.

Mechanical agitation in a water slurry frees a fraction of the unhinged crystals. With present methods such as a mechanical blender, 20 to 30 percent can be released. Since all of the microcrystals are unhinged, this fraction can be increased by improvement in mechanical energy input, preferably by high-shear action.

Typical examples of the preparation of microcrystalline cellulose as described by Battista and Smith (3, 4) are discussed below.

EXAMPLE 2.1 Viscose rayon filaments were subjected to a boiling 2.5 N HCl solution for about 15 min. The residue was separated by filtration and washed with water until neutral. At the end of the washing period the residue was washed with a 1 percent ammonium hydroxide solution to ensure complete neutralization of all acid. The recovered cellulose was then air-dried overnight. It was found to have a level-off degree of polymerization of about 40.

The air-dried product was added to water to form mixtures containing

5, 7, 9, and 10 percent, respectively, of the dried cellulose, and the aggregates in the various mixtures were broken up. Thixotropic gels were formed by subjecting the various mixtures to the action of a Waring Blendor for about 1 h. They were white and highly opaque in appearance.

Small portions of the thixotropic gels were diluted with additional amounts of water while being agitated so as to form suspensions or dispersions containing about 1 percent cellulose. After standing for about 1 h, the larger particles settled into a lower liquid layer, whereas the upper liquid layer assumed a somewhat opalescent appearance. The larger aggregated particles remained suspended in a layer of liquid constituting about one-third of the total volume, without a sharp boundary between the layers. There was a gradual change in appearance from the milklike lower layer to the opalescent upper layer, which constituted about one-half the total volume. Particle-size measurements of the cellulose suspended in the upper opalescent layer were made by visible-light microscopy and indicated that the maximum particle size of the aggregated microcrystals did not exceed about 1 μm, with many particles being less than about 0.5 μm.

EXAMPLE 2.2 Cotton-linters pulp of normal viscose grade was treated with a boiling 2.5 N HCl solution for 15 min. Water was added, and the liquid was stirred and filtered to recover the cellulose, which was then washed with additional quantities of water until substantially neutral; washing was completed with a 1 percent aqueous solution of ammonium hydroxide. The cellulose was then vacuum-dried for about 16 h at 60°C with a vacuum of 29 in Hg. The level-off degree of polymerization was found to be about 200.

A thixotropic-gel dispersion was formed by subjecting the cellulose in a 5 percent concentration in water to the action of a Waring Blendor for about 1 h. More viscous gels were also formed in concentrations up to about 10 percent.

Various process-improvement patents describe the preparation of microcrystalline cellulose and its derivatives (4).

EXAMPLE 2.3 Dissolving-grade wood pulp was treated in a glass-lined digester with 0.5 N HCl at 250°F and approximately 20 lb pressure for 120 min with gentle stirring. Concentration of cellulose solids was 5 percent. The contents of the digester were dumped into a tank and diluted to about 8 percent solids with the addition of water and vigorous stirring. The slurry was subsequently filtered (to about 40 percent solids) and washed to about a pH of 6.5 with water made slightly alkaline (pH 7.5 to 8.0) with dilute NH_4OH. The excess HCl, excess salts, and water-soluble impurities were removed thereby; the resulting filter cake was then reslurried at about 9 percent solids, and the homogeneous aqueous slurry spray-dried to give a white free-flowing powder consisting of completely unhinged or disconnected but aggregated microcrystals. The level-off degree of polymerization as measured by viscosity in cupriethylenediamine solvent was found to be 230.

EXAMPLE 2.4 To produce a grade of dry microcrystalline cellulose which

table 2.2   Bulk Density of Microcrystalline Cellulose from Wood Pulp

	lb/ft^3
Freeze-dried......................................	9.8
2-Propanol-washed	13.3
Methanol-washed................................	14.0
Spray-dried (commercial production)..........	16.0–20.0

is more conveniently redispersed to a stable thixotropic gel, the filtered (40 percent solids) wet cake described in Example 2.3 may be wet-blended with the addition of a hydrophilic *barrier* such as carboxymethyl cellulose (CMC) (approximately 8 to 10 percent) to homogenize the hydrophilic gum throughout the microcrystalline gel system so that the free microcrystals and the aggregates of the microcrystals are coated with a film of the water-dispersible barrier. The homogenized mix is then drum-dried and granulated in a hammer mill or another type of granulating mill into a fine white powder. These special commercial grades of microcrystalline cellulose are readily dispersible in water (28).

PROPERTIES

The bulk density of microcrystalline cellulose dried from various suspending media is given in Table 2.2.

Microcrystalline cellulose is insoluble in water, in dilute acids, and in most organic solvents. It is slightly soluble in sodium hydroxide solution (e.g., 5 percent NaOH).

When dispersed in water, the pH of the supernatant liquid ranges between 5.5 and 7.0.

It retains less moisture than other forms of cellulose under ambient conditions (Table 2.3) and is such a pure form of cellulose that residue of not more than 0.05 percent is found after ignition. Heavy metal content is kept to about 10 parts per million, which again reflects the high level of purity of this form of cellulose (Table 2.4).

A particular advantage of this science over conventional polymer chemistry

table 2.3   Moisture Absorption of Microcrystalline Cellulose from Wood Pulp

% rh (71°F)	H$_2$O,% (after 48 hr)
15	2.5
45	5.7
58	6.0
81	6.2

table 2.4   Chemical and Physical Properties of
Microcrystalline Cellulose from Wood Pulp

Molecular weight	30,000–50,000
Moisture, %	6.5
Organic solvent extractables, %	<0.05
Ash, %	<0.05
Calcium, ppm	<40
Chlorides, ppm	<50
Iron, ppm	<10
Copper, ppm	<4
Solubility:	
Water	Insoluble; dispersible
Dilute alkali	Partially soluble; swells
Dilute acid	Insoluble; resistant
Organic solvents	Insoluble; inert
Oils	Insoluble; inert

(compare Fig. 2.1 and Fig. 2.2) is that by this process not only do you remove, from areas of low lateral order, embedded inorganic impurities, but also you are able to preclude leaving any residual monomer or low-molecular-weight polymer behind in the final product.   This has significance particularly with

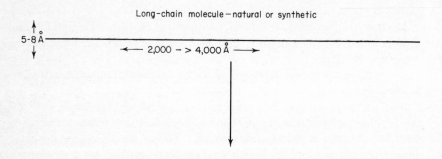

Long-chain molecule – natural or synthetic

5-8 Å

$\longleftarrow$ 2,000 – > 4,000 Å $\longrightarrow$

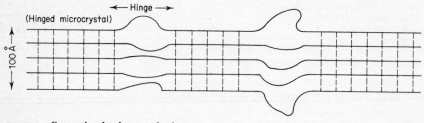

Crystallize single molecules out of solution to form fibers and films

$\longleftarrow$ Hinge $\longrightarrow$

(Hinged microcrystal)

100 Å

fig. 2.1   *Conventional polymer technology.*

Start with natural or synthetic fiber or film-forming polymer

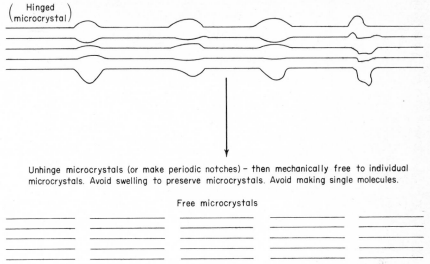

$\left(\begin{matrix}\text{Hinged}\\\text{microcrystal}\end{matrix}\right)$

Unhinge microcrystals (or make periodic notches) - then mechanically free to individual microcrystals. Avoid swelling to preserve microcrystals. Avoid making single molecules.

Free microcrystals

fig. 2.2   *New colloidal microcrystal polymer technology.*

linear synthetic polymer species wherein it is well known that even very small amounts of homopolymer can lead to great variations in the final physical properties of a polymer product.   Of course, the process of unhinging necessarily causes some depolymerization, and so in all cases you end up with microcrystals containing aggregates of molecules whose molecular weight must be less than that of the precursor polymer.

Of course, polymer microcrystals have dimensions that put them into the world of the infinitesimal.   All of our physical measurements of these colloidal systems involve the use of angstroms. Inasmuch as 1 in is 254 million Å, 2.5 million rayon microcrystal particles could be put side by side within the space of 1 in!

A comparison of the size and shape of cellulose microcrystals from different cellulose precursors is helpful in appreciating the unusual properties exhibited by microcrystal polymer suspensoids of celluloses.

First, as illustrated by Fig. 2.3, it is possible to correlate the average length of the constituent molecules in a polymer microcrystal of cellulose, at least, with the actual length of the individual cellulose microcrystals observed directly by electron microscopy (5).

Figure 2.4 shows wood-pulp fibers (from trees in Alaska) whose cellulose microcrystals have been produced by nature, who hooks them together in an orderly lateral geometric form.   Extensive beating of cellulose fibers in water leads to an increasing number of individual fibrils.   It is these fibrils

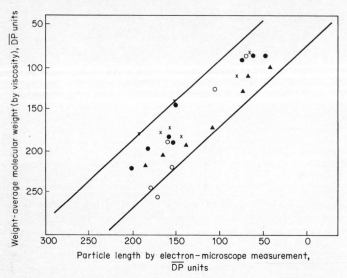

fig. 2.3   *Particle length by electron-microscope measurement, $\overline{DP}$ units (5).*

crossing over each other as they dry down that produce a sheet of paper; the strength of a sheet of paper depends largely on the total number of hydrogen bonds between these interfacial surfaces of fibrils. Yet within each of these fibrils in normal paper are millions of cellulose microcrystals hooked together.

Figure 2.5 is an excellent electron micrograph of a highly beaten paper-making grade of rayon, which reveals that even with regenerated cellulose the opening up of the fibrous structure also leads to the release of long

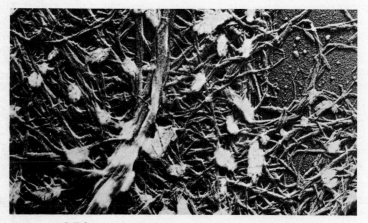

fig. 2.4   *α-Cellulose wood-pulp fibrils.*

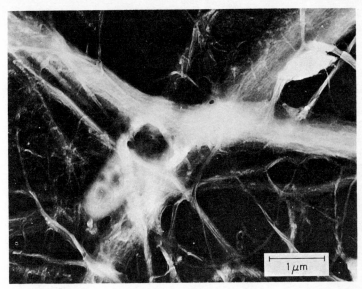

fig. 2.5 *Ultrafine fibrils of a special papermaking grade of rayon, highly beaten.*

ultrathin continuous fibrils. The constituent microcrystals are, of course, still connected together by long-chain molecules interconnecting the microcrystals just as the larger microcrystals are connected in the larger wood-pulp fibrils in Fig. 2.4.

Figure 2.6 shows the single microcrystals being liberated from the matrix of a wood cellulose fiber *after* the molecules between the microcrystals have been severed by the careful use of dilute hydrochloric acid—the hydronium

fig. 2.6 *Microcrystals of cellulose dislodging.*

Atomic model — side view

The cellulose molecular chain — β -1:4 glycosidic linkages between anhydro glucose units

End of chain with a reducing (potential aldehyde) group

End of chain with four reactive OH groups

Atomic model — top view

fig. 2.7 Cellulose molecule.

ion is able to penetrate the hinge regions and break the covalent 1,4-β-glycosidic bonds (Fig. 2.7). However, because the concentration of HCl acid chosen does not swell the microcrystals significantly, the hydronium ion cannot attack the covalent bonds within the matrices of the microcrystals.

Figure 2.8 is a cross section of a single fibril of wood-pulp cellulose which demonstrates clearly the side-by-side packing of the cellulose microcrystals by nature.

As these unhinged microcrystals peel away laterally from each other in this naturally formed matrix, they disperse into the medium (water), and when a sufficient number of these individual microcrystals are dispersed, stable macrocolloidal suspensions result.

Such suspensions produced from microcrystalline cellulose powder may possess a wide range of viscosity properties depending upon the concentration of solids in the suspension as well as on the proportion of free microcrystals to aggregated particles (Fig. 2.9).

Once the individual cellulose microcrystals have been dispersed in water and the resulting dilute stable suspensoid is spray-dried, the microcrystals reaggregate in a random manner (Fig. 2.10) not unlike wooden matchsticks being allowed to fall into a pile on a table (Fig. 2.11).

The random reaggregation of once-liberated polymer microcrystals, how-

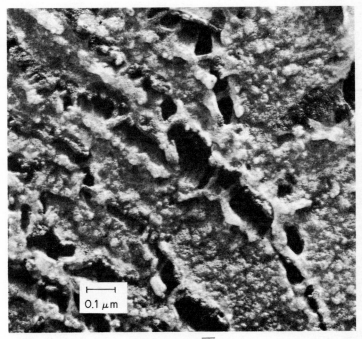

0.1 μm

fig. 2.8   *Cross section of cellulose fibril at $\overline{DP}$.*

fig. 2.9 *Three forms of microcrystalline cellulose.*

ever, necessarily leads to colloidal particle aggregates that are literally *full of holes*. In the case of the wood-pulp microcrystals of Fig. 2.10, the *holes* range from 10 to 100 Å or more. Since the size and distribution of such holes in

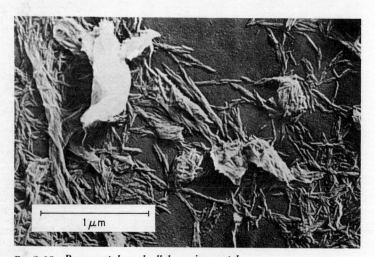

fig. 2.10 *Reaggregated wood-cellulose microcrystals.*

fig. 2.11   Reaggregated matchstick structure simulating reaggregated microcrystals. (The Dow Chemical Co.)

colloidal particles are controllable, the resulting dry powders have commercial utility. (See Chap. 2, Applications.)

Evidence for the internal porosity of the spray-dried microcrystalline cellulose powder is given in Table 2.5 in which the percentage of microcrystalline cellulose needed to dry up or convert many oily and/or sticky products such as cheeses, syrups, etc., to free-flowing forms is listed.

Refractive index data for various forms of microcrystalline cellulose are provided in Table 2.6.

For the new colloidal phenomena to be observed, it is essential that a minimum percentage of the microcrystal particles exist in suspension in unaggregated form, while the remainder may exist in varying degrees of aggregation. Essentially we are concerned in this science with polymer

table 2.5   Granulation Properties of Dry Microcrystalline Cellulose (Conversion of Oily or Sticky Products to Free-flowing Forms)

Ingredient	Commercial microcrystalline cellulose, %
Blue cheese	36.0
Butter	44.8
Cheddar cheese	20.0
Corn oil	39.0
Honey	44.0
Hydrogenated fat	39.0
Lemon oil	50.0
Maple syrup	44.0
Milk chocolate (melted)	32.4
Molasses	44.0
Orange oil	50.0
Peanut butter	23.0
Plastic coconut	20.6
Swiss cheese	13.8

table 2.6 Refractive Indexes of Different Forms of Microcrystalline Cellulose

Sample	Refractive index
1. Wood cellulose	1.556
2. Microcrystalline cellulose from 1	1.576
3. Cellulose microcrystals from 1 dispersed as a gel	1.550
4. Cellulose microcrystals from 1 dispersed as a gel with about 9% CMC	1.556

microcrystals that are in the same dimensional range from that of viruses up to that of the smaller bacteria, roughly 50 to 5,000 Å; at least, that is the preferred colloidal-particle size range. Accordingly, a polymer microcrystal is any particle recovered from a linear natural or synthetic polymer-precursor matrix whose maximum dimension is less than 1 μm and which, in a liquid medium, will form a stable suspensoid. The particles can have different degrees of crystallinity or lateral order.

The size and shape of the microcrystals, as well as the properties of each cellulose suspensoid, depend on the history of the precursor cellulose fibers (Table 2.1).

In Figs. 2.13 and 2.14, respectively, microcrystals recovered from LODP cotton cellulose (Fig. 2.13) are compared with microcrystals recovered from LODP wood cellulose (Fig. 2.14). Cotton microcrystals generally appear thicker and more uniform in length than wood-pulp microcrystals. It is believed that there is considerable significance in the appearance of a "beaded" morphology in the fibrous microcrystals from both cotton and wood celluloses, respectively. Current research is being directed toward elucidating the significance of uniform spherical particles of the order of 100 to 300 Å that may be seen in electron micrographs of each species of polymer micro-

fig. 2.12 Rayon microcrystals.

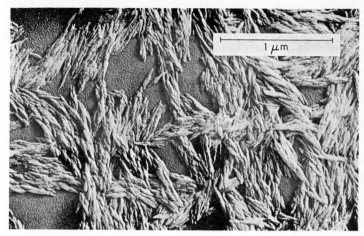

fig. 2.13   *Cotton microcrystals.*

crystals described in this book.   We hope to provide evidence of a uniform concept for explaining the basic morphology of linear high-molecular-weight polymers in a future publication presently under preparation.

Table 2.7 lists size ranges for small unit particles for various forms of cellulose ranging from molecules of derivatives in solution to commercial microcrystalline cellulose particles.   Table 2.8 provides a chart that covers a wide range of the dimensional scale, and it illustrates the narrow range of this scale within which colloidal polymer microcrystals fit.

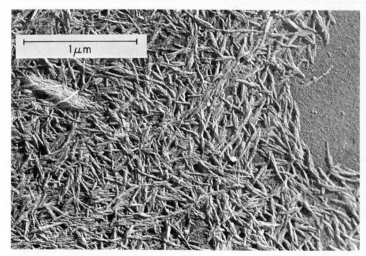

fig. 2.14   *Wood-pulp-cellulose microcrystals.*

table 2.7   Relative Sizes of Particles and Particle Aggregates

Product	Appearance	Range of diameters
Natural and synthetic gums, starches, and water-soluble derivatives	Transparent or translucent aqueous dispersions	5–25 Å
Mechanically disintegrated microcrystalline cellulose	Stable opaque aqueous dispersions	150 Å–5 μm
Pulverized fibrous celluloses	Retains fibrous form; two phases in water	40–500 μm, or higher
Microcrystalline cellulose flour	Fine white powder	Few thousand Å to over 10 μm

As theory would predict, the unhinging of molecules between microcrystals and the dissolution of the soluble degradation fragments leads to an increase in the total crystallinity of the LODP residue. Figure 2.15, an x-ray diffraction pattern of microcrystalline cellulose from wood pulp, clearly confirms this.

Furthermore, conversion of a linear polymer precursor such as wood-pulp cellulose to its LODP sharpens the polymolecularity of the residue as revealed by Fig. 2.16.

If commercial microcrystalline cellulose powder is directly compressed in a mold at room temperature at approximately 20,000 lb/in^2, a dense structural product is obtained having electrical properties, representative data for which are given in Table 2.9.

A dramatic demonstration of the phenomena relating to exposure of new surface forces by liberating *embedded and self-bonded* polymer microcrystals from their precursor matrix is best shown by a product called Avory.

Avory in essence is a *new form of paper*, yet it has all of the outward appearances of *marble* or natural *ivory*. It is produced simply by the air drying (without pressure) of a properly prepared 15 percent suspensoid of wood-pulp cellulose microcrystals.

table 2.8   Pertinent Size Parameters for Colloidal Microcrystalline Polymer Technology

1 in = 254,000,000 Å: single macromolecule is approximately 5 Å in diameter

1 μm = $\dfrac{1}{1,000,000}$ m = 10,000 Å: human hair is about 50 μm (500,000 Å) in diameter

Limit of particle size visible to human eye: $\frac{1}{2}$–1 μm (5,000–10,000 Å)

Size range of viruses and bacteria: maximum dimension is 25–5,000 Å+

Size range of novel polymer microcrystals: maximum dimension is 25–5,000 Å+

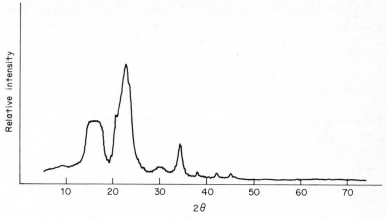

fig. 2.15 *X-ray diffraction pattern of microcrystalline cellulose.*

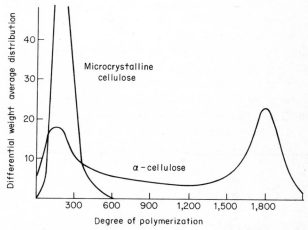

fig. 2.16 *Chain-length distributions (3).*

**table 2.9 *Electrical Properties of Compressed
Microcrystalline Cellulose Powder***

Density, lb/ft^3 86–98
Specific gravity 1.26–1.34
Thermal conductivity, Btu(in)(ft^2)(°F)(h) 1.75
Specific heat, Btu/(lb)(°F) 0.4
Power factor, %(58% rh, 72°F) 2.88
Power factor of vulcanized fiber (electrical grade), % 6.0
Dielectric constant (58% rh, 72°F) 5.6
Impact strength, ft·lb$_f$/in of notch . . . .∼1

Figure 2.6 revealed the dispersion of microcrystals from within their highly ordered matrix in a cellulose fiber. This electron micrograph clearly shows that in this process tremendous new cellulose surface area is becoming available. With it, billions of new surface —OH groups are exposed and ripe for hydration with water molecules.

When a gel of suspended and highly hydrated microcrystals is dried down slowly in air, great internal forces come into play at atmospheric pressure. The tightly bound water is removed slowly, taking days, even weeks compared with overnight, for an equal weight of highly beaten pulp fibers. Ultimately, Avory is produced; a new form of paper (the forces holding the structure together are identical to those in ordinary paper except there are billions more of them) having the outward characteristics of ivory results. Estimates based on the resulting density of Avory structures suggest that internal forces (due to surface tension between the highly hydrated and conformable colloidal microcrystal particles) are of the order of 1.5×10^6 lb/in^2. (The density of Avory may be as high as 1.50. Theoretical density of a single cellulose crystal calculated from x-ray diffraction data is about 1.566.)

Figure 2.17 illustrates one of the unusual properties of this new structural form of cellulose. It shows a block of Avory microcrystalline cellulose produced by drying a gel at room temperature and atmospheric pressure resisting penetration by a 6-in nail.

Figure 2.18 is equally impressive inasmuch as Avory qualifies by definition as being an unusual form of paper. It reveals what happens to two forms of microcrystalline cellulose—one compressed from the powder at high pressure and the other the characteristic ivory-like form dried down from a gel.

Table 2.10 gives a quantitative comparison between the resistance to

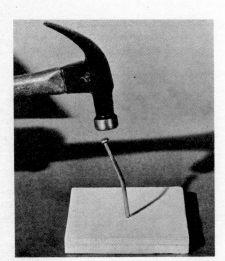

fig. 2.17 *Resistance to impact and penetration by a 6-in nail of a new structural form of microcrystalline cellulose (3).*

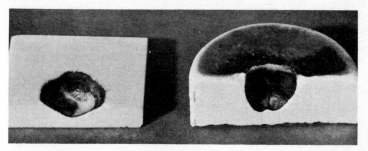

fig. 2.18 *Ablation of dried-gel form and compressed-powder form (3) with oxyacetylene torch (3600 K).*

ablation by an oxyacetylene torch of Avory and that of other common heat-resistant materials.

One explanation of Avory's great resistance is that the temperature of the flame—the energy in the flame—is consumed in breaking billions of hydrogen bonds at the immediate surface of the Avory, which results in a drastic lowering of temperature of the flame at the surface of the Avory sufficient to cause slow carbonization of the durable cellulose surface.

Still another interesting property of high-solids microcrystalline cellulose gels is the ability to transform them by a rolling action into spheres of uniform diameter (Fig. 2.19). When these Avory spheres are carbonized, they also shrink uniformly to about one-third their volume producing ultrapure, hard carbon spheres having properties shown representatively in Table 2.11.

Microcrystalline cellulose suspensoids can be converted into shaped pure-cellulose products which in turn can be converted into structural forms of carbon and/or graphite (Fig. 2.20). These unique new forms of structural paper might well become the preferred raw material from which unusual

table 2.10 High Heat Resistance of Structural Forms of Microcrystalline Celluloses

Product	Bulk density, lb/ft^3	Heat conductance, Btu/(ft^2)(°F)(hr)	Specific heat, Btu/(lb)(°F)	Time under oxyacetylene torch, s*
Compressed flour, ½-in disk	86	1.80	0.4	15 (not through)
Transite, ⅜ in	112	4.50		7 (through; melts)
Marinite, ⅜ in	75–80			7 (through; melts)
Dried gel structure, ¾-in block†	95			70 (crater depth, ½ in)
Steel, ¾ in				5 (through; melts)

*6600°F.
†Dried from 15% microcrystalline cellulose gel.

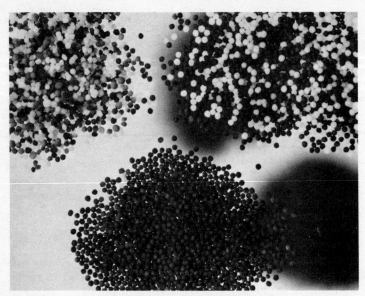

fig. 2.19 *Structural form (spheroidized) of microcrystalline cellulose.*

grades of carbon and graphite may be produced in the future. Laboratory feasibility has been demonstrated.

Extensive studies have been made on the preparation of stable suspensoids and creams from microcrystalline celluloses. The properties of such suspensoids are controlled not only by the history of the precursor form of cellulose but also by the percentage of microcrystals that are deaggregated into single, freely hydratable microcrystals.

Figure 2.21 shows how critical the concentration is during mechanical attrition with a Hobart mixer for building up gel viscosities at a uniform concentration of 15 percent. Data similar to this, but showing varying degrees of efficiency, must be obtained for each specific type of equipment used for mechanical disintegration, although the general trend shown in Fig.

table 2.11 Properties of Carbon Spheres from Microcrystalline Cellulose Spheres

Size (diameter range), mm..........................0.3–0.6
Bulk density, lb/ft³................................. 50
Density (mercury displacement), g/cm³ 1.28
Internal pore volume, cm³/g......................0.052
Surface area (nitrogen method), m²/g...........95
Maximum carbonization temperature, °C 900

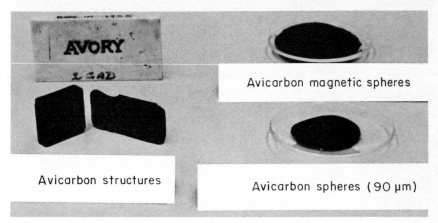

Avicarbon magnetic spheres

Avicarbon structures

Avicarbon spheres (90 μm)

fig. 2.20   *Carbonized forms from microcrystalline cellulose.*

2.21 is valid.   In general, an increase in the viscosity of a gel at a fixed solids concentration relates to the efficiency of microcrystal deaggregation.   As a guideline, the following concentration ranges provide maximum efficiency for shearing away single cellulose microcrystals in an LODP microcrystalline cellulose product: 40 to 50 percent solids for a roll mill, 32 to 36 percent

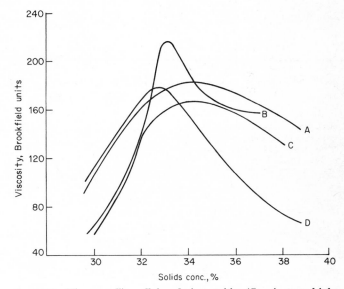

fig. 2.21   *Microcrystalline cellulose during attrition (3).   A, spray-dried, Hobart (commercial); B, never dried, Hobart; C, spray-dried, Hobart (laboratory); D, sample A, Mixmaster.*

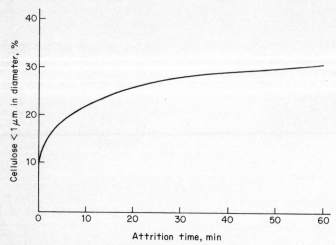

fig. 2.22   *Increase in freed microcrystals with attrition time (3).*

solids for a Hobart mixer with solid paddle, 10 to 15 percent solids for a Waring Blendor or an Osterizer, and 10 to 20 percent solids for piston-type homogenizers.   However, as the curve in Fig. 2.22 shows, once a certain percentage of microcrystals has been freed and hydrated to develop maximum viscosity, it is difficult to produce further deaggregation of the remaining aggregated particles probably because the brush-heap matrix of hydrated single crystals shields the remaining aggregates from direct shear-energy input.

Both pH and the addition of salts can have significant effects on the rheological properties of aqueous microcrystalline cellulose suspensoids.   For example, the maximum apparent viscosity for a 5 percent suspensoid is found at a pH of 10 (Table 2.12).   The changes in the yield stress and shear stress,

table 2.12   Role of pH on Viscosity of 5% Colloidal Dispersions of Microcrystalline Cellulose from Wood Pulp

pH*	Average apparent viscosity†
7	8.8
8	93.4
9	118.2
10	148.8
11	61.1

*By adding NH_4OH.
†Ostwald viscometer, cP.

respectively, on adding 0.08 M NaCl to a 12 percent aqueous suspensoid are given in Table 2.13.

Hermans (23) has carried out a series of investigations in which the flow properties of microcrystalline cellulose aqueous gels were investigated by measuring shear stress as a function of shear rate. Some noteworthy conclusions drawn from his studies are the following:

1. The values of shear stress, at any value of the shear rate, increase greatly with concentration of the gel but do not vary very much with the shear rate, nor do they depend significantly on the flow history of the gel.

2. However, within limits, the yield stress depends on the shear rate applied to the gel prior to the measurement (limited thixotropy).

3. The gels exhibit flow properties that can be interpreted in terms of a particle-network model which is reversibly disrupted by the application of outside mechanical force.

4. During shear the microcrystalline cellulose particles are oriented with the long axes parallel to each other, and so form a liquid crystalline phase.

The typical shear-stress-rate behavior of a series of microcrystalline cellulose gels at varying concentrations is shown in Fig. 2.23; a 3 percent solution of cellulose gum is compared with the microcrystalline cellulose suspensoids and reveals a completely different shear-stress-rate behavior.

A study was made by Edelson and Hermans (24) on the effects of adding electrolyte to microcrystalline cellulose gels. It was concluded that the negative charges present on the cellulose microcrystals do not appreciably contribute to the interparticle forces responsible for the gel properties. Provided the gel and electrolyte solution were properly mixed, only a small increase in the shear stress at any shear rate was measured.

The light-scattering properties, pseudoplasticity, and general viscosity properties of microcrystalline cellulose sols are affected by the size of the

table 2.13   Salt Effects on Avicel Gels: Values of the Two Yield Stresses and the Shear Stress at a Shear Rate of 80 s⁻¹ for 12% Gels in the Presence and Absence of Salt

	Yield stress, dyn/cm^2		Equilibrium shear stress at $80 s^{-1}$
	Low value	High value after shearing at $80 s^{-1}$	
No salt	120	120	1,400
0.08 M NaCl	160	440	2,200
After 1 h at 1,000 s⁻¹, no salt	120	120	1,000
After 1 h at 1,000 s⁻¹, 0.08 M NaCl	160	220	1,400

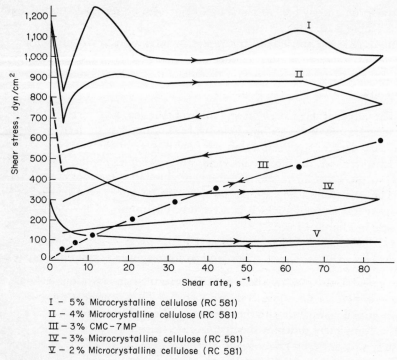

I – 5% Microcrystalline cellulose (RC 581)
II – 4% Microcrystalline cellulose (RC 581)
III – 3% CMC – 7MP
IV – 3% Microcrystalline cellulose (RC 581)
V – 2% Microcrystalline cellulose (RC 581)

fig. 2.23 *Thixotropic and yield values of microcrystalline cellulose suspensoids.*

microcrystals (which varies widely depending on the source), the polydispersity or size distribution (which is influenced by the method and severity of mechanical attrition used), and the total concentration of particles.

Matijevic et al. (26) have made most informative studies of dilute microcrystalline cellulose sols (100 to 400 parts per million) prepared from one grade of commercial microcrystalline cellulose—Avicel PH, a grade made from high-α-cellulose wood pulp. They prepared their sols by diluting an Avicel gel (approximately 10 percent) under constant stirring. The final suspension was subjected to ultrasonic treatment for 90 min. The sediment was separated by decanting, and only the supernatant layer was used in the stability studies. Their results showed that microcrystalline cellulose sols are more sensitive to electrolytes than are typical lyophilic colloids. Coagulation concentrations of simple monovalent counterions are considerably lower than those for many lyophobic colloids. Neither complex counterions nor surface-active ions were capable of reversing the charge of the cellulose microcrystals. This is not altogether surprising since the charge was found to result from the presence of carboxylate groups (—COO—) which were not prone to participate in complex formation.

Table 2.14 lists the values of coagulation concentrations of various counterions for microcrystalline cellulose sols (0.01 to 0.04 percent solids, a pH of approximately 6) as determined by Matijevic et al.

The preparation of smooth, stable, opaque suspensoids of *pure* microcrystals have been described using different precursor cellulose parents. However, a virtually undeveloped process exists whereby specific functional groups (we refer to this as topochemical derivatization of microcrystals) are added to the surface groups of the different cellulose microcrystals (3). This topochemical derivatization must be done with care to preclude overswelling the microcrystals; e.g., very low degrees of substitution must be produced. For example, if a few sodium carboxymethyl groups are added to the surfaces of the microcrystals, gels result which are now translucent instead of opaque; these suspensoids now resemble petrolatum. The sodium carboxymethyl groups on the surfaces of the microcrystals change their electrical surface charge as well as their degree of hydration which in turn effects the bonding between the microcrystals. As a result, high shear energy is not required to produce a suspension of almost 100 percent free microcrystals (Fig. 2.24). The resulting product has properties that suggest it could serve as a greaseless (free of oils or fats) cosmetic base or vehicle. Figure 2.25 shows the appearance of partially oxidized and dispersed cellulose microcrystals.

Topochemical derivatization and its resulting drastic influence on the properties of the other members of the microcrystal polymer family has equally exciting potential and promise though its commercial development lies in the future. This route to readily dispersible novel compositions is quite different from that of making mechanical mixtures of microcrystals and cellulose gums (28).

table 2.14 Coagulation of Microcrystalline Cellulose Sols by Counterions (26)

Counterion	Coagulation concentration, mol/liter
Li^+	1.3×10^{-4}
Na^+	1.3×10^{-4}
K^+	2.0×10^{-4}
Rb^+	2.3×10^{-4}
Cs^+	3.4×10^{-4}
NH^+	3.0×10^{-4}
Tl^{+4}	2.3×10^{-4}
Ag^+	1.0×10^{-4}
Mg^{++}	5×10^{-5}
Ba^{++}	$3-5 \times 10^{-5}$
La^{3+} (pH 5)	8×10^{-6}

fig. 2.24   *Cellulose–CMC microcrystals (topochemical derivatization—degree of substitution of 0.2).*

fig. 2.25   *Oxidized microcrystalline cellulose.*

APPLICATIONS

The commercial applications of microcrystalline celluloses in various physical states are clearly differentiated from the long-standing uses of pure forms of celluloses, which were based almost exclusively on keeping the cellulose in fibrous forms and preserving its molecular weight. In the past every effort was made to avoid degradation of the cellulose to a powdery nonfibrous state. In their nonfibrous and gel forms microcrystalline celluloses have opened up major new outlets for pure celluloses which were never before available for commercial use by man.

The various grades of microcrystalline cellulose have been engineered to contribute unique and useful properties to a wide spectrum of commercial products. For example, the regular pharmaceutical grade (of which Example 2.3 is most representative) serves as a pharmaceutical tablet binder. Millions of pounds of this grade go into the manufacture of tablets annually. It permits the preparation of tableting powder mixes which can be fed directly to a tableting machine, bypassing in many cases the conventional pretableting granulation step.

Other grades (such as Example 2.4) go into frozen desserts to control "heat shock" or ice-crystal growth. Unlike molecularly soluble additives for ice cream, the microcrystalline cellulose particles provide a solid roadblock or barrier to the growth of needlelike ice crystals. Ice crystals nucleate and begin to grow into long needles when frozen desserts are stored; but as they encounter cellulose microcrystals, their linear growth is stopped. Nucleation must start on the other side of the microcrystals. The overall result is that a smaller and more homogeneous size distribution of ice crystals forms which permits production of a smoother, "warmer" frozen dessert. The practical value of this control over ice-crystal growth is that frozen desserts can be inventoried in the fully frozen state over a longer period of time; e.g., they can be manufactured and stored in January for sale in the warm-weather months such as July and August.

Unlike many gel systems based on molecular dispersions of macromolecules, colloidal microcrystalline cellulose gels possess unusually stable viscosities both under ambient conditions and increased temperatures. Even on prolonged standing, the viscosity values of these structural gels remained constant; in other words, microcrystalline cellulose gels do not become rubbery with time as is so characteristic of gels based on the dispersions of hydrophilic macromolecules, such as the various natural and synthetic gums.

Colloidal microcrystalline products are incorporated as rheology-control agents in foods where they may be used as temperature-insensitive thickeners for salad dressings, hollandaise sauces, etc. For example, since gels made from them are substantially viscosity-stable even at temperatures required to pressure-cook foods (250°F), they may be used to produce canned tuna salad, potato salad, etc., in which the salad dressing is sterilized in the can so that such self-contained convenience foods along with the salad dressings are ready for instant use upon opening the can.

Industrial uses of cellulose microcrystals as rheology-control agents are also being developed, although they have not yet received as much commercialization as the well-proven food and pharmaceutical uses.

Nevertheless, they have found applications in water-base paints, decorative laminates, etc. A special grade has been under development for use in ceramics as an additive which aids the green strength of clays, helps moldability, but burns out without leaving a residual color as inorganic additives do. In addition, an industrial grade of microcrystalline cellulose has been found to make possible the continuous coating of resins in the manufacture of decorative laminates and in improved products.

A special grade of microcrystalline cellulose was developed for the pharmaceutical industry where it has found extensive use as a tablet excipient and diluent. It has replaced less expensive competitive products because of its uniqueness. For example, in tableting, starch is a good disintegrant but a poor binder; lactose is a good binder but a poor disintegrant; microcrystalline cellulose, on the other hand, is an excellent binder and a good disintegrant. Pulverized fibrous cellulose, though less expensive than microcrystalline cellulose, is less pure and has poor flow properties. In some pharmaceutical tablet products traces of inorganic impurities can markedly reduce drug stability and shelf life; the unusually low inorganic residues in microcrystalline cellulose (a result of the severe acid digestion used to make microcrystalline cellulose) is a particular advantage for microcrystalline celluloses in pharmaceutical tableting.

The pharmaceutical grade of microcrystalline cellulose is listed in the National Formulary wherein it is described as a fine, white, odorless, crystalline powder, consisting of free-flowing, nonfibrous particles which may be compressed into self-binding tablets which disintegrate rapidly in water.

Food Uses

One of the first major commercial uses for microcrystalline celluloses was as a noncaloric ingredient for engineering the calorie content of foods, especially of fat-loaded foods. In this connection it seems appropriate to describe a key anecdotal incident about the underlying idea that ushered in the creation of this new form of cellulose and one of the first experiments performed to test its applicability as a noncaloric imitation-butter spread.

In 1955, nylon tire yarn was looming as a serious competitor of rayon tire yarn, the latter having all but replaced cotton as a fiber for tire cord. It was well known then that nylon fibers were comprised of very tiny microcrystals that were highly ordered and oriented within the fiber matrix and that these microcrystals were tied together by many molecular hinges—a morphology that helped to explain nylon tire yarn's high strength and toughness especially against impacts. The idea I had at the time was to recover the cellulose microcrystals in rayon tire yarn, disperse them into separate microcrystal particles, and then inject them into the precursor viscose solution. The reasoning was that I could thereby seed the spinning solution so that the rayon yarns produced might be engineered to have even smaller and more oriented crystals than normal and thus to approach more closely the morphology of nylon tire yarns.

This was the original reasoning that led to the design of the first experiment that led to an unexpected result which in turn mushroomed over a period of years into the contents of this treatise!

Starting with LODP cellulose from rayon tire yarn in powder form (a material that had been on the laboratory shelf for many years as part of our fundamental studies of cellulose structure), I asked an assistant, Miss Patricia Smith, to disperse 5 wt percent in distilled water in a laboratory Waring Blendor, suggesting that she subject the mixture to an intense shearing action for at least 30 min. I had hoped that the shearing action would result in infinitesimal crystals becoming suspended in the supernatant water, and it was these crystals or crystal fragments I had hoped to recover in connection with the viscose seeding experiment. Fortunately and fortuitously I had requested a 5 percent dispersion for 30 min; if I had proposed a 1 percent dispersion for 10 min, the completely unexpected result would not have occurred.

Of course, a smooth opaque snow-white suspensoid now characteristic of microcrystalline cellulose gels was produced by the aforementioned Waring Blendor experiment. It was because this gel had the physical characteristics of "colorless butter" and because we knew that there would not be a calorie in a carload of it for human consumption that the potential of this product for food uses was first conceived. As a matter of fact, one of the first products was made by adding carotene, salt, and a synthetic butter flavoring to transform it into a 100 percent calorie-free spread that looked like and tasted like butter. However, when this product was used on hot cakes, it did not melt in the least! The reason, of course, was that the cellulose microcrystals do not melt, and, as will be seen later in connection with commercial applications of this phenomenon, its nonmelting qualities have been put to practical use. We found that the dry powder was most effective for converting oily and sticky food products such as cheese or peanut butter into free-flowing forms (Fig. 2.26).

As an aside, we became so excited by this new gel and the exciting uses it presented to us we never did perform the original experiment that we started out to do. And, many years later, I still urge someone to do it!

Some of the first questions that arose regarding the use of microcrystalline celluloses in foods were: (1) Was it safe to eat as a food ingredient like natural fibrous forms of cellulose were known to be? (2) Did it indeed go through

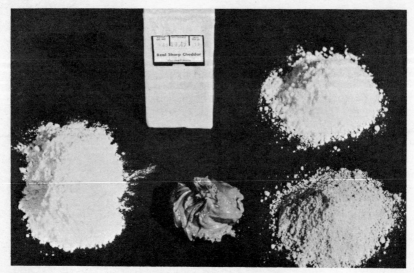

fig. 2.26   *Cheese and peanut butter dried with microcrystalline cellulose.*

the human digestive tract without adding any calories as was well documented for natural fibrous cellulose, nature's normal fibrous bulking agent in more conventional forms of foods?

To this end an interesting microscopical study of the size and shape of both natural fibrous cellulose and microcrystalline cellulose was undertaken in connection with a weight-control experiment using 16 human volunteers under strict medical guidance (29).   This study confirmed that the size and shape of both fibrous particles of cellulose in lettuce, for example, and microcrystalline cellulose particles passed through the human body unchanged (Figs. 2.27 and 2.28).   It was also shown, however, that this conclusion was valid only when human feces containing cellulose were immediately frozen after evacuation and prior to examination as a smear on a microscope slide.   We observed that rapid breakdown of the constituent cellulose particles in human feces occurs at room temperature (Figs. 2.29 and 2.30).

An even more definitive experiment involved preparing microcrystalline cellulose from cotton cellulose which had been recovered from cotton plants growing in a CO_2 environment containing C^{14}.   As a result, some C^{14} atoms were introduced as integral parts of the polyanhydroglucose units within the microcrystals of the cotton fibers (30).   Microcrystalline cellulose labeled with C^{14} was fed to a human subject under appropriate preadaptation periods. No C^{14} activity was observed in the urine or in the exhaled CO_2 of the subject indicating that there was no absorption or digestion of the C^{14}-labeled microcrystalline cellulose.   Analysis of fecal samples showed the recovery

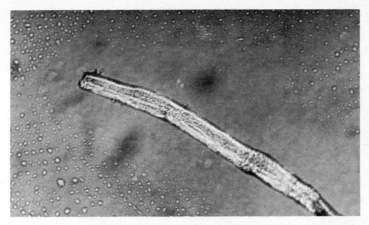

fig. 2.27   Natural fibrous cellulose in human feces—frozen (29).

value of the total administered C^{14} label was 98.9 percent with a standard error of ± 3 percent, which showed that cellulose was not utilized at any significant level.

In addition to the human feeding experiments, extensive skin-patch studies also have shown that microcrystalline products are inert physiologically upon external application.

With the foregoing documentation of the safety and noncaloric efficacy of microcrystalline cellulose established, the development of the unique

fig. 2.28   Microcrystalline cellulose in human feces—frozen (29).

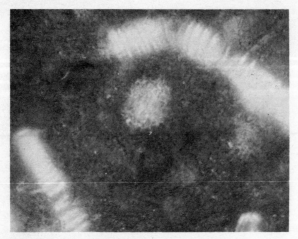

fig. 2.29 *Natural fibrous cellulose in human feces, room temperature—unfrozen (29).*

rheological and other functional properties of oil and/or aqueous suspensoids of microcrystalline celluloses in foods was pursued more aggressively.

Understandably various grades had to be engineered to accommodate special end uses in food as well as in other applications—pharmaceuticals, industrial uses, chromatography, etc.

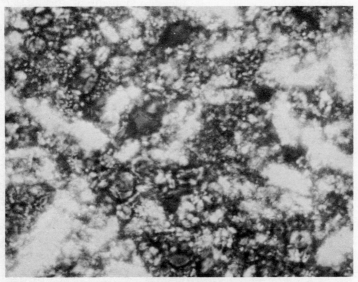

fig. 2.30 *Microcrystalline cellulose in human feces, room temperature—unfrozen (29).*

The most widely used grade of microcrystalline cellulose in the food industry has about 10 percent of medium-viscosity food-grade sodium carboxymethyl cellulose (CMC) blended with it to aid in redispersing the powder with a minimum input of energy. The water-soluble sodium CMC coats the microcrystals and/or aggregates during the spray-drying step in the manufacture of the dry commercial powder, which reduces the overall intensity of hydrogen bonding between the particles as they dry down; in water the molecularly soluble CMC swells rapidly, pushes the particles apart, and leads to a much more rapid formation of a stable suspensoid (28).

Two especially valuable properties are provided by this grade of microcrystalline cellulose: emulsion stability and viscosity stability throughout the thermal cycle. A perfect example of its use is in a line of new cooking sauces wherein it functions as an emulsion stabilizer and thickener at sterilization temperatures.

Suspensoids of microcrystalline cellulose with added CMC have properties quite different from those of the gum solutions, starch gels, or other water-soluble viscosity modifiers. In this respect, the food industry has only recently begun to apply the properties available, which have long been recognized in industrial thickening applications. They are especially valuable because they are unaffected by food acids that degrade starches and gums, at pH values up to about 2.5.

Although they behave like an emulsifier, microcrystalline cellulose products do not lower surface tension in the manner of a conventional surfactant. Microcrystalline cellulose particles are wetted by both oil and water, and the particles locate at the oil-water interface of an emulsion, which produces an interfacial film of great strength and stability. In addition, the particles thicken or gel the water phase. Oil droplets are restricted effectively by this gel to provide the emulsion stability sought.

Microcrystalline celluloses with added CMC are making it possible to produce shelf-stable food products never before possible. The stabilizing ability of microcrystalline cellulose will permit formulating novel spreads, toppings, sauces, casseroles, convenience entrees, as well as sterilizing flavored products normally sensitive to heat processing. In addition, formulations meeting commercial standards have been perfected using a sterilizable *carrier* emulsion to which meat, fish, or poultry can be added, making possible retail packs of refrigerated-type products normally found only as delicatessen items. Similar techniques allow the production of meat or fish spreads, unusual dairy-based ingredients for toppings, sour cream, low-calorie dressings, nondairy coffee creamer, imitation processed-cheese spread, or heat-stable cream sauces, all having long-shelf-life and controllable-viscosity advantages.

Typical properties of a microcrystalline cellulose–CMC grade are given in Table 2.15.

table 2.15 Typical Properties of Microcrystalline
Cellulose-CMC, Food-Grade

	Grade 581	Grade 501
Composition, parts:		
Microcrystalline cellulose	89	91.5
Sodium carboxymethyl cellulose	11 ± 1	8.5 ± 1
Physical form	White, water-dispersible powder	
Particle size	Averages 35 μm, which is 50% less than 400 mesh; when dispersed, particle size reaches 0.15 μm	
Moisture level, %:	6 or less when shipped, hydroscopic	
Content @ 25% rh	5.0	
@ 50% rh	7.0	
@ 75% rh	10.2	
Bulk density, lb/ft^3	31	
Ash, %	2	
Heavy metals, ppm	10	
Iron, ppm	5	
pH:		
2% dispersion (aqueous)	6-8	
1% dispersion (aqueous)	6-8	
Solubility	Insoluble in water and organic solvents; disperses in water to form colloidal sols and gels	
Viscosity, cP:		
1.2% concentration	120 ± 40	
2.1% concentration		120 ± 40

Specially formulated mixes comprising microcrystalline cellulose have been developed for frozen-dessert products—ice creams, ice milks, and other dairy product–based frozen desserts. The tiny, noncaloric, discrete cellulose microcrystal particles are effective at relatively low concentrations for controlling the heat-shock phenomena associated with storing frozen desserts, such as ice milk and ice cream, for long periods of time. The ice crystals that do grow remain smaller, which leads to a smoother, richer-tasting frozen dessert, a practical benefit for both the manufacturer and the consumer.

Microcrystalline cellulose dispersions containing CMC are highly thixotropic and exhibit finite yield values due to the brush-heap matrix developed by the solid microcrystals in suspension touching each other. These suspensions lose viscosity rapidly on being stirred and upon resting their yield value increases quickly to an equilibrium value.

Pharmaceutical Uses

The largest single original commercial use for microcrystalline cellulose was in the tableting of pharmaceuticals. It is now used for this purpose in every major country of the free world. Microcrystalline cellulose performs as an excipient to assist in the flow, lubrication, and bonding properties of the ingredients to be tableted; to improve the stability of the drugs in tablet form; and especially to give rapid disintegration of the tablet in the stomach of the user to ensure quick availability of the drug. (See Fig. 2.31.)

There are three commercial methods for preparing pharmaceutical tablets: (1) wet granulation, (2) dry granulation, often referred to as *slugging*, and (3) direct compression. The advantage of direct compression is obvious. Microcrystalline cellulose can be used in the direct compression of most drugs (estimates run from 60 to 80 percent), and because of the savings in capital, equipment, and labor these advantages offset the higher price of micro-crystalline cellulose giving a net savings in tableting costs.

Shangraw et al. (31) investigated microcrystalline cellulose in tableting, and their conclusions clearly summarize the reasons why this form of cellulose has become so widely used.

Tablets consisting of pure microcrystalline cellulose were first compressed on a hand-operated Stokes Model A-3 single-punch press using $\frac{3}{8}$-in flat-faced punches and then on a Colton Model 216 rotary press using $\frac{7}{16}$-in standard concave tooling. The following characteristics were noted:

1. Compression on both machines was excellent. Extremely hard tablets could be made with ease without evidence of overloading the presses.

2. There were no signs of lubrication difficulties even after compressing 1,000 tablets on the single punch.

3. Flow properties of the material itself were good, but the addition of 1 percent Cab-O-Sil pyrogenic silica did increase the weight slightly and reduce the weight variation indicating more uniform die fill.

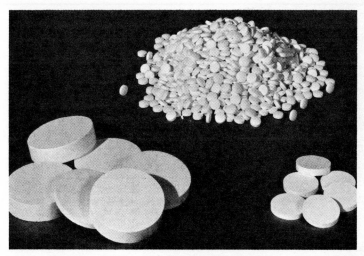

fig. 2.31   *Tablets of compressed microcrystalline cellulose.*

4. Disintegration times on tablets compressed to a Strong–Cobb hardness of 4.5 to 10 were so rapid that they could not be accurately measured in a USP disintegration apparatus. As harder tablets were compressed, disintegration times could be measured.

5. Tablets exhibited excellent friability. One batch of tablets produced on the rotary press with an average weight of 0.990 g and a Strong–Cobb hardness of 7 showed a 0.35 percent loss when tumbled in a Roche Friabilator for 500 revolutions at a speed of 23 r/min. It is pointed out that, although the hardness is relatively low and friability conditions extreme, little chipping was obtained. Loss of weight during friability tests occurs as a result of erosion on tablet faces rather than edges.

Industrial Uses

Technical grades of microcrystalline celluloses are available for a wide range of industrial uses.

INDUSTRIAL USES IN WHITE POWDER FORM

Function	Use
Reinforcing filler	Thermosetting resins
	Thermosetting laminates
Tableting binder and disintegrant	Detergents
	Disks for active carbon preparation
Flow control	Powder mixtures
Color vehicle	Dye and pigment transport
Chromatographic separation	Thin-layer chromatography
	Column chromatography
Cellulose raw material	Controlled-particle-size cellulose derivatives

INDUSTRIAL USES OF COLLOIDALLY DISPERSED SUSPENSOID FORMS

Products	Functional roles
Wax-in-water emulsion	Sole emulsifier
	Shear stability
	Freeze-thaw stability
	pH stability
	Aqueous-solvent miscibility
	Thixotropy
Latex emulsions	Emulsifier
	Stabilizer
	Thixotropy
	Prevents "strike through"
	Adhesion
Water-base-paint formulation	Flow control
	Brushing control
	Viscosity control
	Bubble leveling
	Sagging control

Paint stripper.............................. Thixotropy
                                            Versatile emulsifier
                                            Viscosity control
Herbicide emulsion .... Sole emulsifier
                                            Stabilizer
                                            Thixotropy
                                            Aqueous dispersibility
Ceramics .... Plasticity
                                            Modulus of rupture
                                            White firing
                                            Nonmigrating binder
                                            Rapid prefired glaze hardening
                                            Nonfoaming glaze suspension

Thin-layer chromatography (TLC) is an important laboratory method in the separation, purification, and identification of chemical mixtures. It combines speed with the sensitivity, flexibility, and precision of paper and column chromatography; routine separations are performed in minutes rather than hours.

Microcrystalline cellulose substrates are receiving increasing attention as binder-free substrates for chromatoplates, and the related published scientific literature is growing significantly. Battista et al. performed the original work on the use of microcrystalline cellulose in chromatography (32–35). Many other investigators have researched the use of microcrystalline celluloses for chromatography, and an excellent compilation of these references is available (36).

Pure nonfibrous microcrystalline cellulose with an average particle size of 50 μm forms chromatoplates superior to other chromedia with respect to coating quality and resolution.

Wolfrom et al. (37–39) performed separations of a representative range of organic and inorganic compounds using microcrystalline cellulose extrusion-column chromatography. The technique correlated with chromatoplate separations. One- and two-dimensional separations were performed. In many cases R_f values were identical to those reported for paper chromatography. Compact spot size resulted in easy compound differentiation.

Preparation of Chromatoplates Using Microcrystalline Cellulose Binder-free chromatoplates are prepared from aqueous slurries containing 15 to 30 wt percent of microcrystalline cellulose. Optimum coating quality and resolution are assured when 15 to 20 percent aqueous slurries are blended for approximately 1 min with a Waring Blendor or laboratory mixer. Increased blending improves coating smoothness, but the time must be limited since a gel will form with prolonged blending. The mixture is spread on the desired-size plates, which are then dried, preferably in air at room temperature. However, they may also be dried with care at elevated temperatures.

This product adapts to all TLC techniques and apparatus, forms hard and durable chromatoplates, and can be mass-produced and stored for future use with no loss of qualitative properties.

In addition, colloidal forms of microcrystalline cellulose have found uses in water-base paints, especially in those which contain hydroxyethyl cellulose; a synergistic effect has been observed leading to improvements in the properties of some dripless paints at lower costs (40).

In ceramic glazes, colloidal microcrystalline celluloses act as nonfoaming aids, whereby the outer coatings are more uniform and the drying times are shortened (41).

Applications in the areas of sizing of fibers, waterless hand cleaners, household polishes and cleaners, paint strippers, paper coatings, and leather finishes are being evaluated and results are promising.

One of the more interesting industrial applications of microcrystalline cellulose was developed by the Asahi Chemical Company. High-solids microcrystalline cellulose suspensoids may be converted on drying under the proper tumbling conditions into hard *ivory-like* spheres. The diameters of these smooth spheres can be regulated, and the incorporation of a wide range of ingredients within them has been projected. [See Chap. 2, Properties (Avory) and also Battista et al. (42–45).]

Inasmuch as the surface-tension forces and the coincident hydrogen-bonding forces are as powerful as they are, when water is removed from an aqueous microcrystalline cellulose suspensoid, the microcrystalline gels may be used, after inorganic fillers are suspended in them, to produce hard hybrid spheres. Spherical catalyst carriers always have been sought after by industry.

Heavy loadings of an inorganic catalyst support material such as γ-alumina, or silica, for example, are possible. Once the hybrid suspensoid is converted into perfect spheres using microcrystalline cellulose binding forces, the organic microcrystals of cellulose may literally be burned away. This leaves a highly porous γ-alumina sphere having minimal friability. Typical diameters of the macropores in such spheres range from 3,000 to 12,000 Å. The uniqueness of this ingenious route to the engineering of controlled macropores in ceramic catalyst structures such as catalyst supports lies in the versatility it affords for producing macropores of almost any size and distribution. Advantages of the microcrystalline cellulose route as an adjunct in producing novel and superior catalyst supports and substrates for many of the process industries are:

1. Spheroidization of various types of inorganic catalysts or their carriers is possible in high yields and is more economical than it was before.

2. Pore volume and an increase in active sites can be regulated simultaneously.

3. Macropores can be produced with controlled diameters and a homogeneous diameter-size distribution.

4. The macropores are determined by the size of the microcrystalline cellulose particles which are ultimately burned away.

5. The advantages of surface area, packing, and handling of smooth spherical catalyst substrates are well known in various process industries that are catalyst dependent.

CHAPTER THREE

Microcrystalline Collagens

INTRODUCTION

Paralleling cellulose, which is the most abundant structural polymer in the plant world, collagen is the most abundant fibrous protein polymer in mammals and as such plays a structural role of great biological significance. It has been estimated that collagen comprises between 30 and 60 percent of the total protein content of mammals and up to 30 percent of the total content of organic material. The primary locations of the protein are skin, bone, and tendon, and on a dry basis, the corium layer of skin; i.e., the dermis consists of 90 to 95 percent collagen. However, collagen fibers are not confined to these regions of cellular organization but can be found in practically every tissue and organ.

The Oxford Dictionary of 1893 defined collagen as "that constituent of connective tissue which yields gelatin on boiling," and this reflects the etymological sense. The word collagen is taken from the Greek *kolla* meaning glue and *egenomen* meaning to produce. However, this definition is hardly adequate today, and the term may be defined more precisely by saying that collagen, irrespective of the source from which it is obtained, exhibits the following characteristics:

1. One in three of the amino acids are glycine residues and about one in five are imino acid residues.

2. There is a characteristic wide-angle x-ray diffraction pattern with a meridional arc at 2.86 Å and hydration-sensitive equatorial reflections or spots at 11 Å.

3. A striated structure may be observed in the electron microscope with a periodicity of 640 Å. This characteristic spacing is also demonstrated by small-angle x-ray studies.

4. The infrared absorption spectrum possesses a peak at 3330 cm^{-1} which corresponds to the —NH stretching frequency. This shows a 30 cm^{-1} displacement from the 3300 cm^{-1} peak which is typical of most proteins.

5. There is a specific rotation of about $-350°$ in solution which falls to about $-120°$ upon heating above a certain transition temperature.

6. In the solid state a shrinkage phenomenon is exhibited upon heating. The x-ray pattern becomes diffuse, i.e., that of an amorphous solid above the shrinkage temperature.

Nature has produced in collagen a polymer architecture of remarkable sophistication, an "engineering" achievement that qualifies collagen for a role of far greater versatility and complexity than any other known man-made or natural high-molecular-weight polymer.

An anecdotal reflection on nature's morphological masterpiece in collagen is contained in the enormously high *wet* tensile strength of collagen fibers. For example, fibers of bovine tendon collagen have tensile strengths (dry basis) of as high as 70,000 lb/in^2, or about the same as typical piano wire! And, despite this these fibers are extremely pliable and biologically inert.

It now generally is agreed that the native collagen fiber is built up by a highly ordered process of linear and lateral aggregation of thin, highly elongated chains of amino acids into a discrete basic molecular entity called tropocollagen. In other words, collagen has a multiphase morphology comprising varying degrees of aggregation (lateral order) of constituent tropocollagen molecules superimposed upon the intramolecular structure and intermolecular bonding of the long-chain–three-strand helical amino acid molecules within the tropocollagen macromolecular unit particle. A further distinguishing feature of collagen's fundamental chemical structure is the unique high glycine (about 30 percent) and proline and hydroxyproline (imino acids) content of the amino acid chains; no other class of proteins possesses such high levels of these imino acid molecular building units. It is also worth noting that the sulfur amino acid content of collagen is low, and disulfide bonds are not available for tertiary structure formation and stabilization. The high imino acid content also precludes the establishment of an intramolecular hydrogen-bond-stabilized secondary structure such as the α-helix as there will be no hydrogen atom available at peptide bonds involving proline or hydroxyproline.

The common basic unit or building block of collagen, tropocollagen, is an asymmetric, rodlike macromolecule with dimensions of about 3,000 $\times$ 14 Å and a weight-average molecular weight of about 300,000. The preparation of soluble collagen involves the separation and isolation of tropocollagen from predominantly insoluble collagen precursor materials.

Tropocollagen comprises three separate polypeptide chains—one α_2 chain and two α_1 chains coiled along a three-fold–left-handed screw axis. In addition, the axes of the individual chains are twisted into a right-hand superhelix, and so a rotation of 324° is required for one turn of the helix in individual chains, with a translation of 8.58 Å. As there is a complement of three amino acids per turn, the translation between individual amino acids is therefore 2.86 Å. The three chains are arranged such that their axes are parallel and would appear to form the corners of an equilateral triangle with sides of approximately 5 Å if viewed along the axis of the superhelix. The structure is stabilized by interchain hydrogen bonds between adjacent carbonyl and amide groups and also by fewer but stronger covalent bonds.

Throughout most of the repetitive amino acid sequence every third residue is glycine, which is followed commonly by proline and often preceded by hydroxyproline. This regularity is missing only at both ends, which are also of particular interest because it is here that the covalent cross-links occur. Large parts of both the α_1 and α_2 chains have been sequenced, and the various tracts are indexed along the chains in terms of the fragments that are generated by cyanogen bromide cleavage at the few methionine residues. The α_1 chain gives rise to nine and the α_2 to six such fragments.

Collagen exhibits still other unusual characteristics: in mammals, including man, it possesses a closely defined melting point slightly above the normal body temperature for each species, and in vivo it possesses remarkable metabolic inertness!

Starting with a highly refined grade of native bovine corium collagen, our studies since 1964 have been directed to the conversion of it into a *microcrystalline* form—not unlike the aggregates of "limiting microfibrils" proposed by Veis et al. [(46), p. 933].

Native bovine corium collagen exhibits, as would be expected, a much lower lateral order (lower overall level of crystallinity) than does cellulose inasmuch as collagen as produced in nature undoubtedly comprises a mixture of species that are different constitutionally. This overall low level of lateral order for intact collagen is understandable. Nevertheless, native collagen does exhibit a definitive wide-angle x-ray diffraction pattern. Conversion of native bovine corium collagen into a (precursor) morphology that can be mechanically disintegrated into discrete colloidal *fragments* comprising aggre-

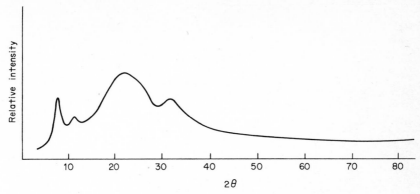

fig. 3.1 *Never-dried bovine corium collagen, air-dried. X-ray diffraction pattern.*

gations of tropocollagen molecules laterally bonded essentially as produced in nature requires a careful chemical pretreatment. This pretreatment must maintain degradation at a minimum and also preclude molecular dissolution of the constituent tropocollagen molecules. Enzymes and/or other chemical pretreatments (such as salt solutions) are *not* used. After extensive cold-water extraction at an approximately neutral pH, swelling is carefully controlled by means of selective exposure to ethyl alcohol–water extractions during which a partial acid salt (normally the HCl partial acid salt although many other partial acid salts have been made) is formed. These steps produce a product containing precisely the number of milliequivalents of HCl salt formation such that, when it is dispersed in water and mechanically attrited, a uniform pH of 3.2 ± 0.2 is maintained. Furthermore, such suspensoids maintain their very high viscosities (e.g., at only 0.5 percent solids the viscosity is in excess of 12,000 cP) at room temperature (25 °C) for long periods of time. The product after conversion to the partial acid salt does exhibit a slightly sharper wide-angle x-ray diffraction pattern than the native corium collagen precursor from which it is produced. Complete loss of intensity at approximately 10.8 Å occurs when the pH is lowered and degradation of the microcrystalline collagen is allowed to proceed to a level approaching gelatin (Figs. 3.1 to 3.3).

PREPARATION

Some details of the procedures for the preparation of microcrystalline collagens remain unpublishable for proprietary reasons. However, the published patent literature pertaining thereto is becoming increasingly informative and information contained in this section has been gleaned from these published references (47–54).

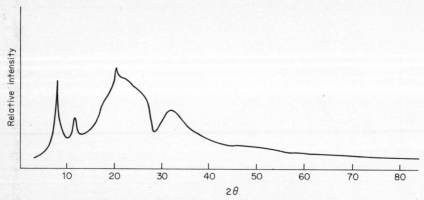

fig. 3.2   *Microcrystalline collagen, air-dried.   X-ray diffraction pattern.*

An overview of the process for conversion of bovine corium collagen into the microcrystalline precursor state as herein defined follows. The method for producing a water-insoluble, ionizable, partial salt of collagen consists of distributing through a body of undenatured, fibrous, natural collagen an aqueous solution of an ionizable acid having a pH between about 1.6 and 2.6 based upon a 1 percent solids concentration. The acid is allowed to react with the available amino groups of the collagen to form a water-insoluble, ionizable, partial salt of collagen containing between about 0.4 to about 0.7 mmol of acid calculated as HCl/g of collagen based upon collagen containing approximately 0.78 mmol of primary amino groups/g of collagen. The temperature is maintained below 30°C, and the partial salt as recovered is essentially free of individual tropocollagen molecules and degraded derivatives thereof.

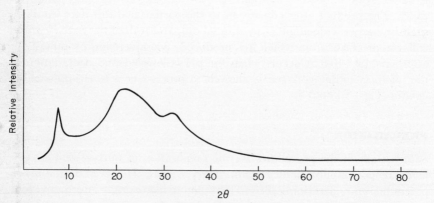

fig. 3.3   *Highly degraded microcrystalline collagen (air-dried, gelatin equivalent).   X-ray diffraction pattern.*

More specifically, this method involves the following sequential steps:

1. Recovery of a special grade of native green corium from healthy steers by splitting the freshly removed and thoroughly washed hides below the hair-follicle line and above the flesh-side line so as to recover the pure sandwich of corium collagen between these two outer surfaces

2. Extensive washing, purification, and inspection of the corium splits so obtained

3. Mechanical comminution of the purified, wet-backed corium splits (about 25 percent solids) into a *diced* form so that they can be more readily handled

4. Controlled swelling of these particulate pieces of corium collagen in ethyl alcohol–water media, involving intermittent centrifuging

5. Formation of the partial acid salt (usually HCl) whereby the milliequivalents of bound HCl are precisely controlled

6. Appropriate drying of the partial-acid-amine-salt cake and its subsequent conversion to a definitive "flour" physical form and packaging

Table 3.1 provides actual data showing the relationship between milliequivalents of HCl, collagen concentration, and actual pH's of suspensoids prepared from the corresponding partial HCl amine salt.

table 3.1   Milliequivalents of HCl / g of Collagen vs. Gel pH

HCl, meq	Collagen, g	meq/g	Gel pH
	1% solids		
4.0	4	1.0	—
3.18	4	0.795	2.83
2.525	4	0.632	3.15
2.0	4	0.5	3.41
	0.75% solids		
3.0	3	1.0	2.7
2.385	3	0.795	2.87
1.90	3	0.633	3.18
1.55	3	0.516	3.49
1.192	3	0.397	3.75
0.948	3	0.316	4.04
	0.5% solids		
1.262	2	0.6305	3.27
1.002	2	0.502	3.40
0.796	2	0.498	3.64

Microcrystalline Collagen
Hemostat Adhesive

The formation of the partial acid salt of purified bovine corium collagen flour (Fig. 3.4) for use as a unique hemostat adhesive (54,57) involves the careful addition of 0.58 mmol of HCl to a limited proportion of available NH_2 groups. This partial salt formation builds into the natural fibrils the precontrolled amount of HCl which, *if and when the fibrils are subsequently dispersed in water at 0.5 percent solids using high mechanical shear,* leads to a hydrolysis of the constituent (salt) HCl molecules leading to an aqueous suspension which equilibrates at a pH of 3.2 $\pm$ 0.2.

In the case of the microcrystalline collagen hemostat flour it is essential that the pretreated partial acid salt in natural fibrillar form is never dispersed in water so that visibly the original fibril morphology remains unchanged. This is precisely paralleled with cellulose fibers, which remain physically intact prior to dispersing precursor microcrystalline cellulose in water to release microcrystals and form stable aqueous suspensoids (3).

Specific examples of the preparation of microcrystalline collagen are taken from the patent literature (49). Microcrystalline collagen may be defined as follows:

> A composition comprising a water-insoluble, ionizable, partial salt of collagen having a bound ionizable acid content of from about 0.4 to about 0.7 millimole of acid calculated as HCl per gram of collagen based upon collagen containing approximately 0.78 millimole of primary amino groups per gram of collagen, being essentially free of individual tropocollagen molecules and degraded derivatives thereof and being further characterized in that when colloidally dispersed in water to form a one-half percent by weight gel wherein at least 10 percent by weight of the partial salt has a particle size under 1 micron, the gel exhibits a pH of about 3.2 $\pm$ 0.2 and exhibits an essentially constant viscosity after about 1 hour for at least 100 hours when stored in a closed container at 5°C.

fig. 3.4   *Microcrystalline collagen flour.*

EXAMPLE 3.1 Ground vacuum freeze-dried cowhide is soaked in an aqueous hydrochloric acid solution having a pH of 2.0 (2 g ground vacuum freeze-dried hide, 200 ml dilute acid) for 15 min at room temperature and then treated in a Waring Blendor for 25 min at a temperature not higher than 25°C; about 16 percent of the solids are of submicron size. The resulting gel has a pH of 3.0 and an apparent viscosity, measured at 25°C on a Brookfield viscometer with TB spindle at 10 r/min, of 42,900 cP.

Using an acid solution having a pH of 3.0, the product has a pH of 5.4, exhibits little bodying action (apparent viscosity under same conditions 4,800 cP), and is replete with undispersed fibers. A stable dispersion is not formed even after prolonged mechanical disintegration. At a pH of 2.6, an apparent viscosity of about 30,000 cP is obtained. The viscosities of products rise to a peak when the pH of the treating acid solutions is in the pH range 2.0 to 2.3 and drops off rapidly such that below a pH of 1.7, the viscosity is again below 30,000 cP apparent viscosity, reflecting degradation of the collagen and the formation of soluble lower-molecular-weight components. Using an aqueous hydrochloric acid solution having a pH of 1.3, a large proportion of the hide substance is degraded to very low-molecular-weight material. The disintegrated product has a pH of 1.5 and a viscosity of less than 100 cP.

The quantity of acid as set forth in this example is based upon treatment at a 1 percent solids concentration or consistency. As the concentration of collagen solids is increased, corresponding increases in the quantity of acid will be required to achieve the preferred final pH range of the dispersions.

The amount of submicron material may be measured by sedimentation, after dilution of an aqueous gel to a sufficiently low concentration so that the liquid is sufficiently thin to allow heavy particles to separate out. Specifically, a filtered 1 percent gel prepared as above is diluted by at least a factor of 10. This is allowed to stand 6 h, and the percentage of solids in the top fifth of the material is measured. This material, under the microscope, indicates it to be substantially all of submicron size. The percentage of submicron material in the total sample is calculated from the amount found in the top fifth. It is not essential that this procedure be followed precisely in measuring the content of submicron material. So long as microscopic examination of the top aliquot reveals the absence of substantially all material above one micron, the method may be used. It is highly desirable to pass the suspensoid through a fine-pore filter prior to dilution.

EXAMPLE 3.2 Ground vacuum freeze-dried cowhide is used to prepare 1 percent solids dispersions or gels by the same method as described in Example 3.1 using hydrochloric acid solutions of different concentrations. The initial pH of each solution and the pH of the final dispersions are measured (Table 3.2). The viscosities of the dispersions are determined as described in Example 3.1, and visual observations are also noted.

table 3.2   Criticality of pH to Control Gel Viscosity

Sample	Initial pH	Final pH	Viscosity, cP	Visual observations
A	3.12	5.41	4,800	Undispersed fibers
B	2.39	3.69	38,400	Highly viscous, transparent gel
C	2.37	3.58	43,200	Highly viscous, transparent gel
D	2.07	2.83	42,900	Highly viscous, transparent gel
E	1.42	1.50	3,840	Low viscosity, opaque gel
F	1.13	1.18	4,800	Low viscosity, opaque gel
G	0.43	0.45	1,600	Low viscosity, opaque gel, presence of oil phase

Preparation of Microcrystalline Collagen Suspensoids

Translucent viscous, time-stable thixotropic, pseudoplastic suspensoids showing a definite yield point can be readily prepared by dispersing microcrystalline collagen in water.   High shear is necessary to produce such suspensoids. The Waring Blendor and Cowles Dissolver are suitable for low concentrations and sigma blade, Hobart mixers, or the Rietz Extructor for high concentrations.   Maximum viscosity is obtained in the pH range 3.0 to 3.4.   At low pH's (2.5) hydrolysis and disintegration of the tropocollagen entity occurs and viscosity decreases; the rate and extent increase with time, temperature, and decreasing pH.   Above pH 4, the structure of the suspensoid is very heterogeneous and collagen fibers begin to coagulate.   Salts and ethanol or isopropanol also coacervate gels at certain concentrations.

Nevertheless, a unique feature of microcrystalline collagen suspensoids is their compatibility in the presence of a wide range of hydrophilic components; e.g., methanol, glycerin, ethylene, propylene and butylene glycols, formamide, dimethyl sulfoxide, and ethanol- or isopropanol-H_2O solutions up to about 40 percent alcohol/60 percent H_2O.   Microcrystalline collagen will also gel a variety of foodstuffs—vinegar, citrate juices, sugar solutions, cheese spreads, ketchup, mustard, gelatin, and sorbitol solutions.   The suspensoids also are excellent suspending agents for particulate matter due to the brush-heap matrix of colloidal particles touching at high-enough solids content.

Filtration of aqueous suspensoid gels removes small amounts of unreacted fibers and fibril fragments.   Films cast from filtered gels dry to clear, slightly anisotropic films.   Cross-linking with formaldehyde, glutaraldehyde, alum, and other tanning agents is feasible.   Films adhere well to many substrates—cotton, paper, rayon, wool, glass, aluminum, silk, and nylon. Mylar and Dacron polyester surfaces and Teflon are excellent substrates for casting films as they offer ready release for their removal.

Laboratory Preparation of Microcrystalline Collagen Gels

From Nonpreheated Flour Suspensoids of the desired concentration are prepared by attriting fibers in the desired medium for 15 to 30 min in a Waring Blendor. The optimum lab-scale gel quantity is 400 g. Concentration should not vary more than ± 0.01 percent for reasonable viscosity reproducibility up to solids contents of 1 percent.

It is recommended that the microcrystalline collagen be allowed to soak for 5 min in the medium and that this be followed by attrition for 2 min at a Powerstat setting of 20 and for 15 min minimum at a setting of 60 to 110 (depending on viscosity) to maintain a controllable vortex. These conditions are applicable to systems having viscosities up to 12,000 to 16,000 cP at 25°C. During attrition, the temperature of the mixture is maintained at 25°C by surrounding the jar with a plastic bag partially filled with dry ice and by reducing or completely stopping agitation intermittently. Sides of the jar are policed at intervals with a rubber spatula to ensure more homogeneous mixing.

From Heat-pretreated Flour (122°C, 30 h)

1. Precool distilled water to a 4°C (refrigerator) temperature.

2. Using a standard-size (1 liter) Waring Blendor, add 12 g of sterile flour to 400 ml of precooled water.

3. Run the Waring Blendor at high speed for at least 30 min or until an extremely viscous homogeneous suspensoid is produced. The temperature of the mixture should be maintained at 25°C or less during this treatment.

4. Dilution of an aliquot of this gel to 0.1 percent solids or less with distilled water at a pH of 3.2 ± 0.2 will provide a stable suspensoid in which a large percentage of *microcrystals* are suspended, a suspensoid from which an aliquot may be dried down for electron microscopy, etc.

Preparation of Partial Salts of Different Acids

Other acids may be substituted for hydrochloric acid to form corresponding partial acid salts. In such instances, the concentrations of acids and the ranges must be varied. Furthermore, it has been found when plotting gel viscosity obtainable against hydrogen-ion concentration in the acid solution, the peaks obtained with other acids may be sharper, which indicates a much narrower useful concentration and greater difficulty in preventing degradation and buildup of acid-soluble tropocollagen. Sulfuric acid gives such a low, sharp peak that its use in commercial operations is counterindicated; the peak is so narrow that control is much too difficult. Acetic acid, hydrobromic acid, and cyanoacetic acid all give high peaks but much narrower peaks than hydrochloric acid, and so greater care must be exercised in their use; even when they produce high-viscosity gels, variations in localized concentration during treatment can produce sufficient localized conversion to tropocollagen to make dry products which lose their integrity when immersed in water. From a practical commercial viewpoint and in view of its low cost and the absence of complications in its use, hydrochloric acid is the treating agent of choice for most end uses.

The optimum viscosity of the gels obtained at different specific pH values when different acids are utilized is illustrated by the following example:

When gels containing 1 percent solids were prepared from ground vacuum freeze-dried cowhide and acid solutions had different pH values as described in Example 3.1, the initial pH of each solution and the pH of each final dispersion or gel were measured. The viscosities of the dispersions also were determined as described in Example 3.1. The results of these determinations are given in Table 3.3.

table 3.3

Acid	Initial pH	Final pH	Viscosity, cP
HCl	3.12	5.41	4,800
	2.39	3.69	38,400
	2.37	3.58	43,200
	2.07	2.83	42,900
	1.42	1.50	3,840
HBr	3.00	4.60	4,000
	2.38	3.67	44,300
	2.06	2.57	28,800
	1.35	1.42	7,200
H_2SO_4	3.02	4.53	4,000
	2.34	3.45	25,700
	2.05	2.55	19,700
	1.50	1.60	6,000
CH_3COOH	2.98	3.51	39,400
	2.37	2.58	28,000
	2.06	2.25	18,000
	1.42	1.58	1,600
$CNCH_2COOH$	3.00		43,200
	2.56		
	2.24		49,600
	2.15		37,000

It will be noted from the foregoing data that the specific maximum viscosity of the gels varies with the different acids. When these and other data are plotted using log-log coordinates, it is found that, in general, maximum viscosities are obtained when the acid solutions used have an initial pH of about 2.3 ± 0.1 and the pH of the gels is about 3.2 ± 0.2.

Preparation of Stable Freeze-Dried Microcrystalline Collagen Structures

Twenty grams of chopped-up cowhide, free of water by freeze drying, is placed in 1,980 ml of a hydrochloric acid solution having a pH of 2 and treated at 25 to 30°C in a Cowles Dissolver, Model IVG, for 15 min at 5,400 r/min using a 4-in pick blade. At the end of the attrition, the 1 percent gel of microcrystalline colloidal collagen is spread in a freeze-drying tray to form a layer 1 in thick, and freeze-dried overnight (-40 to -50°C, vacuum 5 μm, heating cycle not exceeding 30°C with condensation of sublimed water at 60°C). The resultant product is a $\frac{3}{4}$-in mat which absorbs 65 times its own weight of water. The tensile strength of a dry test strip 1 in in width is $3\frac{1}{2}$ lb, and the wet strength of a like test strip is quite low, but measurable.

Test samples of the mat can be repeatedly immersed in water, air-dried, and reimmersed in water without disintegrating. Upon immersing the mat in water, it swells but retains its integrity even when allowed to remain in the water over periods of years. This action of the dried product illustrates that the original macromolecular morphology of the microcrystalline colloidal particles has been retained to an extent sufficient to preclude disintegration to a true molecular dispersion of tropocollagen. In other words, the partial salt of collagen contains sufficient natural bonds that hold the original collagen molecules together laterally to render the product water-insoluble in the presence of the bound acid.

The retention of a portion of the original bonds is also evident from a consideration of the bound acid in the dry products. When the amino acid residues of bovine corium collagen, for example, are considered, 1 g of collagen contains approximately 0.78 mmol of primary amino groups available to react with an added acid. Actual analyses of products derived from microcrystalline colloidal collagen gels prepared with various acids show a bound-acid content varying from about 0.4 to about 0.7 mmol of acid (calculated as HCl)/g of collagen with an average bound-acid content of about 0.58 mmol of acid/g of collagen. Accordingly, it is concluded that certain of the amino groups are bound in the inner region of the microcrystalline particles and are unavailable to react with the acid in view of the controlled swelling conditions under which the partial acid salt is initially formed.

PROPERTIES

Figure 3.4 illustrates the basic physical form of the bovine-microcrystalline-collagen raw material—a fluffy fibrous flour form. Essentially most known end uses, with the exception of the important hemostat-adhesive product in

table 3.4 Typical Composition for Native Bovine Corium Collagens

	%
N_2	17.0–18.0
Hydroxyproline (N-base)	6–9
Proline (N-base)	10–14
Hydroxylysine (N-base)	0.8–1.2
Glycine (N-base)	24–30
Albumins	2–5
Lipids (varies with season and age)	1–7
Reticulin	<0.4
Carbohydrates as hexosamine	<0.60
Elastin (appearance—characteristic yellow fibers)	<1.0
Glycoprotein	<0.5
Mineral components	0.75–1.5
Collagen per se	85–92
(specific *wide*-angle x-ray spacing, and *small*-angle scattering shows periodicity banding of 640 Å)	

which it is used per se in sterile form, require conversion of this form into suspensoids or structural forms.

Table 3.4 gives representative analytical data for thoroughly water-washed native bovine corium collagen.

These data may be compared with those given in Table 3.6, which lists a typical analysis for medical-grade microcrystalline flour prepared from native bovine corium. These data illustrate clearly the unusually high purity of this form of native collagen with its nitrogen content approaching the theoretical value for pure collagen and its noncollagenous organic and inorganic impurities reduced to extremely low levels. The uniqueness of microcrystalline collagen is that it is a select partial acid salt which provides inherent automatic control of the aqueous pH as well as dry storage stability at room temperature.

A representative analysis of a natural catgut collagen suture is listed in Table 3.5. Table 3.6 provides comprehensive chemical, physical, and microbiological properties for medical-grade microcrystalline collagen.

table 3.5 Typical Analysis of Commercial Natural Collagen Suture

Tyrosine, mg/g	4.85
Tryptophan, mg/g	0.281
Proline, mg/g	99.8
Hydroxyproline, %	9.87
Hexosamine, mg % as glucosamine	17.4
Uronic acid, mg/g as glucuronic acid	22.2

table 3.6 *Typical Analysis of Medical-Grade Microcrystalline Collagen*

Raw material parameter	Typical value
Water-holding capacity, g H_2O/g	.27
Moisture (oven wt loss),* %	$.10 \pm 2$
Bulk density, lb/ft^3	2.5
Refractive index (flour form)	1.544–1.548
Viscosity (Brookfield), cP (0.5% aqueous gel) (24 h)	25,450
pH at 25°C (0.5% aqueous gel)	3.2 ± 0.2
Refractive index (0.5% aqueous gel)	1.332–1.334
Refractive index (clear film)	1.540
Color (prior to sterilization)	White
Form	Fibrous
Electrokinetic potential	Positive
Odor	Slight proteinaceous
Heavy metals, ppm (total)	<10
Arsenic, ppm	<0.2
Lead, ppm	0.4
Antimony, ppm	<2.0
Ether extractables, %	<0.30
Nitrogen, % (approaches theoretical)	17.4–18.2
Total chlorine, % (as is basis)	2.5
Total ash, %	0.03
Residual ethanol, %	0.1

*2–3 h (110°C).

MICROBIOLOGY

Bacteriological Examination[a]	
Determination	Microcrystalline collagen flour, organisms per gram
Total plate count	<200
Coliforms	nd[b]
Staphylococci	nd[b]
Anaerobic sporeformers (Clostridia)	nd[c]
Salmonella[a,d]	nd[e]

[a] "Recommended Methods for the Microbiological Examination of Foods," 2d ed., American Public Health Association, 1966.

[b] nd = nondetected at a sensitivity level of 5 organisms per g.

[c] nd = nondetected at a sensitivity level of 1 organism per g.

[d] "Bacteriological Analytical Manual," Section 14, U. S. Department of Health, Education, and Welfare, Food and Drug Administration, April 1, 1966.

[e] nd = nondetected at a sensitivity level of less than 20 organisms per g.

An analysis of gelatin which is produced from collagen by degradation and depolymerization provides (Table 3.7) an excellent breakdown of the amounts and types of amino and imino acids nature uses in constructing natural collagen.

Table 3.8 provides extensive side-by-side analyses of the amino acids found in several forms of native bovine corium collagen, microcrystalline collagen as defined in this treatise, and a commercial cross-linked form of gelatin.

A parallel Table 3.9 of data by I. V. Yannas (55) of the amino acid profile of human tendon collagen is also included for comparison purposes. The consistencies of the amino acid profiles of various collagens and gelatins are noteworthy.

One may better understand the engineering ingenuity which nature has developed to produce the magnificent morphology of the polymer we call collagen by examining the numerous bands in its IR spectrum as reported also by I. V. Yannas (55). These bands are listed in Table 3.10.

Microcrystalline collagen flour per se or in converted forms normally possesses relatively low bulk densities (Table 3.11).

Microcrystalline collagen even in small amounts builds viscosity in glycerin and glycerin-water mixtures. For example, whereas the viscosity of

table 3.7 Gelatin Amino Acids—Averages

Type of amino acid	Nature of R in the general formula $NH_2CH(R)COOH$	Name of amino acid	Approximate g amino acid/100 g dry protein
	—H	Glycine	25.5
	$—CH_3$	Alanine	8.7
	$—CH_2OH$	Serine	3.3
Aliphatic	$—CH(CH_3)_2$	Valine	2.9
	$—CH(OH)·CH_3$	Threonine	2.0
	$—CH_2·CH·(CH_3)_2$	Isoleucine	
	$—CH(CH_3)·CH_2·CH_3$	Leucine	7.1
Aromatic	$—CH_2·C_6H_5$	Phenylalanine	3.7
	$—CH_2·C_6H_4OH(1,4)$	Tyrosine	1.0
Sulfur-	$—CH_2·S·S·CH_2—$	Cystine	0.2
containing	$—CH_2·CH_2·S·CH_3$	Methionine	0.9
	$—CH_2·C:CH·NH·C_6H_4(1,2)$	Tryptophan	0
Heterocyclic	$—CH_2·CH_2·CH_2$-(cyclic)	Proline	19.7
	$—CH_2·CH(OH)·CH_2$-(cyclic)	Hydroxyproline	14.4
Acid	$—CH_2·COOH$	Aspartic acid	5.6
	$—CH_2·CH_2·COOH$	Glutamic acid	11.2
	$—CH_2·C:CH·NH·CH:N$	Histidine	0.8
Basic	$—(CH_2)_3NH·C(:NH)·NH_2$	Arginine	8.7
	$—(CH_2)_4·NH_2$	Lysine	5.9

table 3.8   Amino Acid Profile of Bovine-based Collagen Products

Amino acids, mg/g	Native bovine (steer) corium	Delimed bovine corium, food-grade	Microcrystalline collagen*	Gelfoam† (cross-linked gelatin)
Lysine	32.9	39.0	36.0	35.1
Histidine	7.65	9.16	7.61	7.36
Ammonia	3.82	4.61	4.01	4.03
Arginine	72.6	90.9	80.7	79.5
Aspartic acid	52.1	60.6	63.7	57.4
Threonine	15.9	18.8	18.8	17.0
Serine	28.5	34.6	33.9	32.0
Glutamic acid	90.6	108	106	103
Proline	93.3	111	114	113
Glycine	204	248	238	229
Alanine	86.8	106	97.2	96.5
Cystine	nc‡	nc	nc	nc
Valine	19.8	23.8	22.7	18.0
Methionine	6.92	7.11	6.18	5.64
Isoleucine	13.7	16.5	16.1	12.8
Leucine	27.1	31.8	31.5	29.6
Tyrosine	7.42	8.72	8.12	7.66
Phenylalanine	17.8	21.3	20.8	20.5
Tryptophan	0.300	0.279	0.322	0.358
Hydroxyproline	103.0	122.0	137.0	114.0

*Product of Avicon, Inc.
†Product of Upjohn Co.
‡nc means not calculable.

essentially 100 percent glycerin is about 600 cP, a 99.5/0.5 glycerin–microcrystalline collagen dispersion is as high as 46,000 cP, over a sevenfold increase in viscosity (Table 3.12).

The partial hydrochloric acid salt of collagen in the microcrystalline form is unusually stable. It has been stored at ambient temperatures in excess of 3 years without exhibiting any significant change in 0.5 percent–suspensoid viscosity. This unique stability permits the use of dry heat to sterilize microcrystalline collagen in its fluffy flour form. Changes in surface area for two lots before and after dry-heat treatment in an oven for 28 h at 120°C are shown in Table 3.13. A comparison of changes in the bulk density and gel pH before and after the same heat pretreatment is made in Table 3.14.

Dry-heat treatment of microcrystalline collagen at 120°C for 28 h leads to partially irreversible hydrogen bonding, which reflects itself in a sharply reduced apparent viscosity of the suspensoid, as well as in a highly coalesced microcrystal (Fig. 3.27).

If one tries to make a suspensoid of a 0.5 percent solids using heat-treated

table 3.9 The Amino Acid Composition of Collagen (Human Tendon) (55)*

Trivial name	R in repeating sequence —NHCHRCO—; or imino acid formula	No. of amino acid residues/1,000 total residues
Alanine	$-CH_3$	110.7
Glycine	$-H$	324
Valine	$-CH-CH_3$ $\quad\ \ CH_3$	25.4
Leucine	$-CH_2-CH\begin{smallmatrix}CH_3\\CH_3\end{smallmatrix}$	26.0
Isoleucine	$-CH-CH_2-CH_3$ $\quad\ \ CH_3$	11.1
Proline	CH_2-CH_2 $CH_2\ \ CH-COOH$ $\quad\ \ NH$	126.4
Phenylalanine	$-CH_2-\bigcirc$	14.2
Tyrosine	$-CH_2-\bigcirc-OH$	3.6
Serine	$-CH_2-OH$	36.9
Threonine	$-CH-OH$ $\quad\ CH_3$	18.5
Methionine	$-CH_2-CH_2-SCH_3$	5.7
Arginine	$-CH_2-CH_2-CH_2-NH-C\begin{smallmatrix}NH_2\\NH\end{smallmatrix}$	49.0
Histidine	$-CH_2-C-N$ $\quad\ HC\ \ CH$ $\quad\quad\ NH$	5.4
Lysine	$-CH_2-CH_2-CH_2-CH_2-NH_2$	21.6
Aspartic acid	$-CH_2-COOH$	48.4
Glutamic acid	$-CH_2-CH_2-COOH$	72.3
Hydroxyproline	$HO-CH-CH_2$ $\quad\ \ CH_2CH-COOH$ $\quad\quad\ \ NH$	92.1
Hydroxylysine	$-CH_2-CH_2-CH-CH_2-NH_2$ $\quad\quad\quad\quad OH$	8.9

*P. 67.

table 3.10 Major Bands in the IR Spectrum of Collagen (55)*

Frequency, cm^{-1}	Interpretation
4850–4890	Combination of N—H stretch (3300 cm^{-1}) and in-plane C—N—H deformation (1550 cm^{-1})
4580–4600	Combination of C=O modes
3450–3400	Free water
3290–3330	N—H stretch
3060–3100	Overtone of Amide II
2930–2950	C—H stretch
2870	C—H stretch
2850	C—H stretch
1710	Nonionized COOH
1640–1660	C=O stretch (Amide I)
1560	COO$^-$ band
1535–1550	N—H in-plane deformation plus C—N stretch (Amide II)
1445–1455	CH$_2$ deformation and CH$_3$ asymmetric deformation
1407	COO$^-$ band
1375–1391	CH$_3$ symmetric deformation
1310–1340	CH$_2$ wagging
1230–1270	C—N stretch plus N—H in-plane deformation (Amide III)
1075–1082	CO vibrations of hydroxyl groups
940	O—H deformation of COOH (?)
920	O—H deformation of COOH (?)
650	N—H deformation, out-of-plane

*P. 68.

table 3.11 Bulk Densities of Various Collagen Products

Sample	Bulk density, lb/ft^3
1. Gelfoam (Upjohn)	0.6–1.0
2. Braun (German) collagen implant pad	4–5
3. Microcrystalline collagen flour	2.5–6.0
4. Freeze-dried microcrystalline collagen pad (from 2.5% solids)	1.75–2.00
5. Freeze-dried microcrystalline collagen pad (from 5% solids)	3.50–4.50
6. Microcrystalline collagen in nonwoven web forms	10–18

table 3.12   Viscosity of Microcrystalline Collagen in Glycerin and Water

| Sample | wt % | | | pH | Viscosity* (25°), cP |
	Avitene	Water	Glycerol		
A	0.5	99.5	0.0	3.2	16,280
B	0.5	74.5	25.0	3.0	21,160
C	0.5	49.5	50.0	3.0	24,000
D	0.5	24.5	75.0	2.9	36,400
E	0.5	0.0	99.5	2.9	86,400

*After 25 min in Waring Blendor.

table 3.13   Surface Area of Microcrystalline Collagen before and after Treatment at 120°C for 28 h

| Sample | Surface area, m²/g | |
	Before	After
Lot A	9.1	3.2
Lot B	8.3	2.5

table 3.14   Physical Properties of Heat-treated Microcrystalline Collagen Samples

| Property | Untreated | | Heat-treated (28 h at 120°C) | |
	A	B	A	B
Color	White	White	Off-white	Off-white
Bulk density, lb/ft³	2.10	1.75	2.10	1.79
Gel pH (24 h)	3.2	3.2	3.0	3.0

microcrystalline collagen following the standard procedure for preparing such a gel from the non-heat-treated flour, a two-phase mixture is obtained. The heat-treated form rehydrates slowly and only partially. However, as was the case with microcrystalline celluloses, stable suspensions of heated microcrystalline collagen may also be prepared by attriting the product in water at much higher-solids contents. For example, 3 percent solids or higher, using vigorous agitation, leads to stable, thick suspensions which may then be diluted to thinner gels as needed. (See procedure, p. 67.)

Mechanical shearing of any microcrystalline product at high solids (5 to 40 percent), especially if the microcrystals are aggregated by means of hydrogen bonding, always leads to a more efficient deaggregation of the individual colloidal particles and to more stable aqueous suspensoids than can be obtained by the same method of attrition of a low-solids mixture.

Morphological Studies of Bovine Collagen and Microcrystalline Collagen Produced Therefrom

Extensive studies have been undertaken by us, beginning in 1964, using electron microscopy, electron-scanning microscopy, and optical microscopy to study the morphology of bovine corium collagen and products derived from it. The following series of photographs represent some of our more interesting results.

Figure 3.5 is a phase photograph of bovine corium fibers wet-backed in water. Figure 3.6 is an electron micrograph of a single fibril of bovine corium collagen showing the characteristic 650- to 700-Å bond periodicity for an unswollen collagen fibril.

Exposure of bovine corium collagen fibers to 200 proof alcohol followed by freeze drying leads to a denatured state as shown in Fig. 3.7. Complete loss of periodicity occurs as would be expected.

Figure 3.8 shows the interesting phenomena observed when unswollen collagen fibers were treated as a suspension in a Waring Blendor and they were allowed to wrap around themselves at certain times leading clearly to visible cross-over effects. Another heterogeneous effect observed with unswollen but beaten corium fibrils is presented in Figs. 3.9 and 3.10, wherein a "banana peeling" of fibrils from within a larger fibril appears to occur.

Figures 3.11 and 3.12 are characteristic of the ballooning of a swollen collagen fibril in a strongly acid medium. This effect seems to parallel the ballooning phenomena extensively observed for cellulose fibrils swelling in

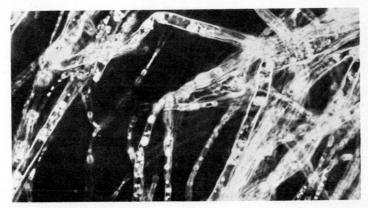

fig. 3.5   Collagen fibers (phase).

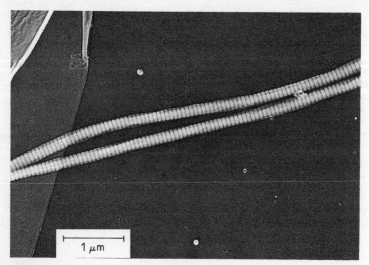

fig. 3.6   *Unswollen bovine corium collagen fibrils.*

cuprammonium hydroxide solution.   We are unable to provide a definitive interpretation of these scattered and truly heterogeneous observations at this time.

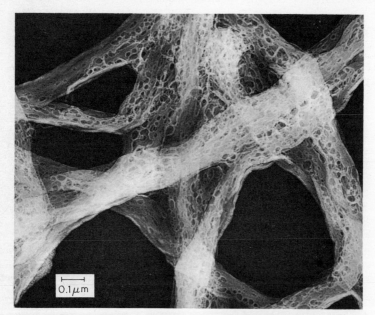

fig. 3.7   *Denatured bovine collagen fibrils after treatment with 200 proof ethyl alcohol, and freeze-dried.*

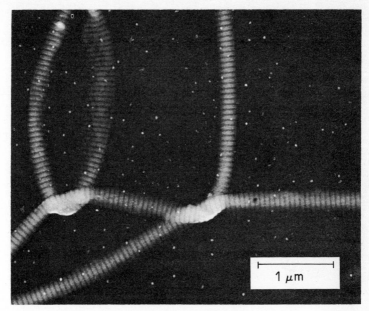

fig. 3.8 *Bovine corium collagen fibrils showing wrap-around.*

When purified (by extensive cold-water extraction at a pH of about 8) corium fibrils are allowed to swell in an aqueous medium at a pH of 2.5 or less, even at room temperature, the fibrils appear to unwind and swell up extensively with complete loss of the banded periodicity. Prolonged standing at this pH at room temperature (or more rapidly at higher temperature) causes a progressive swelling (Figs. 3.13 and 3.14) and concomitant degradation.

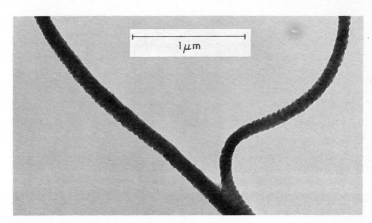

fig. 3.9 *Unswollen bovine collagen fibrils separating.*

fig. 3.10 Bovine fibrils showing "banana peel" effect.

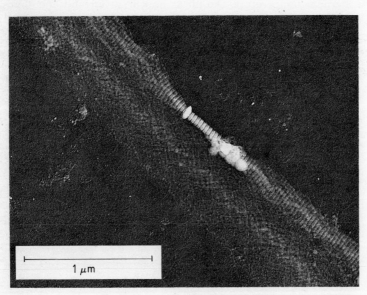

fig. 3.11 Bovine collagen fibril showing "ballooning" effect.

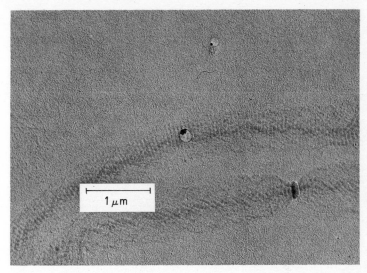

fig. 3.12 Bovine collagen fibril showing extension of "ballooning" effect.

If bovine corium collagen fibrils (without prior organic-solvent extraction) are swollen directly by aqueous hydrochloric acid at a pH of as low as 2.0 and then dispersed in a Waring Blendor, extremely small globules of fat are found to be interdispersed intimately among or within the swelling fibrils (Fig. 3.15).

As bovine collagen is swollen in acid, it progressively loses its gross fibril morphology and disperses first to the basic molecular building block of

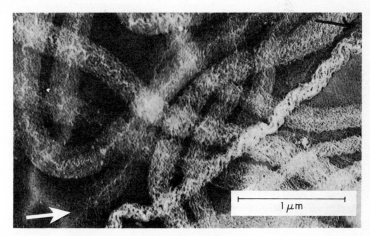

fig. 3.13 Highly swollen bovine collagen fibrils.

fig. 3.14   *Ultrafine structure of swollen bovine collagen fibrils.*

collagen known as tropocollagen shown clearly in Fig. 3.16, an electron micrograph by Hall (56).

The beaded morphology of the tropocollagen molecules is worthy of particular note. This structure is interpreted as the regular interlooping of tropocollagen's three constituent amino acid chains (two α_1 and one α_2), each with limited degrees of freedom to rotate due to interlocking lateral bonding forces.

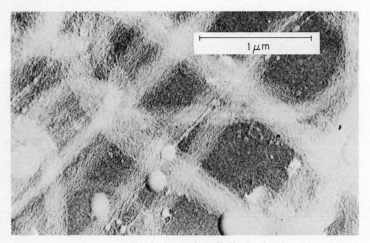

fig. 3.15   *Unextracted swollen bovine corium fibrils showing fat globules.*

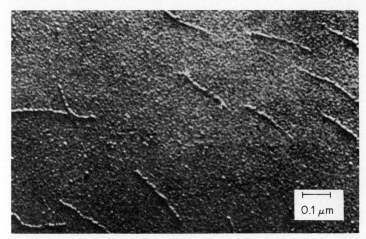

fig. 3.16 *Tropocollagen molecules* [*Hall, C. E. (56)*].

In this regard the recovery of tiny microcrystal spheres from the heat-pretreated partial acid salt of bovine collagen as shown in Fig. 3.17*a* has been intriguing. The size of these tiny microcrystal spheres almost coincides with the size of the beads that appear in the tropocollagen molecules.

In the course of our studies, we have observed consistently submicron particles of the same general size and shape recoverable from other polymer

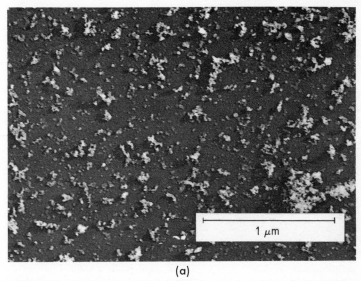

(a)

fig. 3.17(a) *Smallest observed microcrystals of bovine collagen (partial acid salt) after heat treatment for 30 h at 122°C.*

precursors following conversion to the microcrystalline state and severe mechanical attrition at 5 percent or higher solids. For example, Fig. 3.17*b* illustrates such particles recovered from fibers of natural silk. Figure 3.17*c* reveals particles of similar size and shape recovered from wool fibers. Figure 3.17*d* shows particles obtained after treating wood pulp microcrystals at 5

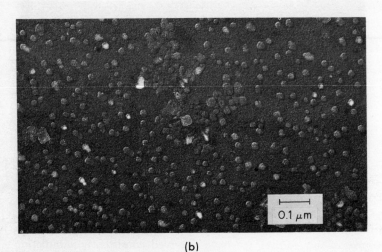

(b)

fig. 3.17(b)   *Microcrystals from natural silk.*

(c)

fig. 3.17(c)   *Microcrystals of wool (note precursor fibrils).*

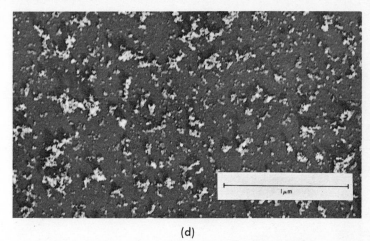

(d)

fig. 3.17(d)   *Wood-pulp-cellulose microcrystals (produced by treating microcrystalline cellulose 100 min in an Osterizer at 5 percent solids).*

percent solids in water with prolonged severe mechanical shear. We have assembled in the Appendix a complete series of electron micrographs of all such particles in the 100- to 300-Å size range recovered from a wide spectrum of natural and synthetic polymers.

It is our experience that the fat content can be reduced to a very low level in natural undenatured bovine collagen by extraction with appropriate organic-solvent–water mixtures while the temperature and the degree of swelling of the fibrils also is carefully controlled. Formation of specific partial acid salts when such restricted conditions of swelling are enforced makes it possible to engineer into the collagen pH control within narrow limits, for example, easily to within ± 0.2 pH units. Through such careful pH control an environment of controlled swelling is provided whereby the original natural morphology is preserved to a degree so that subsequent mechanical attrition permits the conversion of the native partially swollen fibers into discrete fibril fragments.

Ultimately, in the case of collagen, at pH's less than about 2.6, there is swelling at the molecular level such that the tropocollagen triple helix is opened up and degradation of the constituent amino acid chains takes place, which leads to a complete loss of the characteristic beaded structure for tropocollagen and the formation of a form of gelatin (Fig. 3.18). In Fig. 3.18 one can see remnants of the precursor tropocollagen molecules embedded in the filmlike morphology of the gelatin.

We recognize, of course, that collagen is the least crystalline (in the strict x-ray crystallographic sense) of the polymers studied, exhibiting as it does a relatively diffuse x-ray diffraction pattern. It is for this reason that the microcrystalline partial-acid-salt form exhibits only a relatively small sharp-

fig. 3.18   USP gelatin—aqueous solution dried to a film.

ening of the x-ray diffraction bands, and the discrete microcrystal particles do not reveal a high degree of structure or lateral order in the electron micrographs.   This is as would be expected from theory.   The x-ray diffraction patterns for the microcrystals recovered from other polymers that are substantially more crystalline than collagen to begin with overwhelmingly confirm the predictions of theory

Figure 3.19 shows a fibril of bovine corium collagen in a controlled degree of swelling as the partial acid salt.   The application of high-speed mechanical

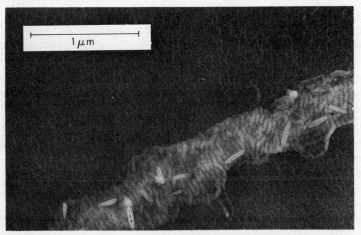

fig. 3.19   Bovine collagen fibril undergoing controlled swelling as partial HCl salt (microcrystal precursor).

shear to the fibrils in this state leads to the fibril fragments or microcrystals shown in Figs. 3.20 and 3.21. High-pressure filtration of these microcrystals causes them to spherulize as shown in Fig. 3.22a. The size of these micro-particles depends largely on the pretreatment conditions of the collagen fibers as well as on their concentration during mechanical shearing. If mechanical shearing is done at 5 percent solids in an Osterizer at 25°C for 60 min, unit particles shown in Fig. 3.22b are observed in the supernatant liquid of a

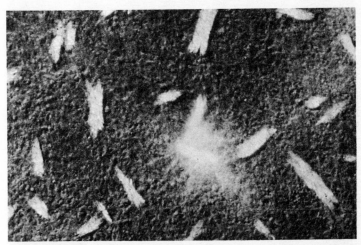

fig. 3.20 *Microcrystals or "microfibrils" of bovine collagen partial acid (HCl) salt (recovered from supernatant of a diluted suspension made by treatment in a Waring Blendor for 20 min at high speed at 25°C).*

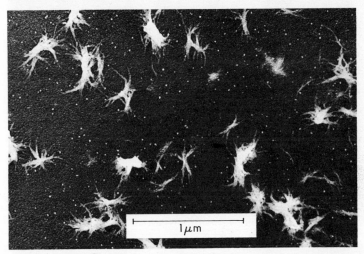

fig. 3.21 *Microcrystals of partial acid salt of bovine collagen after severe mechanical shearing [20 percent solids (aqueous) through Rietz Extructor].*

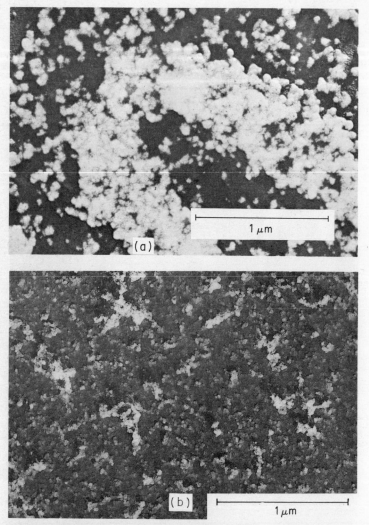

fig. 3.22   (a) *Microcrystals of partial acid salt of bovine collagen filtered through fine glass filter prior to heat treatment.* (b) *Bovine collagen microcrystals prior to heat treatment after 60 min at 5 percent solids at high speed in an Osterizer (supernatant particles of solution diluted to 1 percent solids).*

diluted suspension.   In addition, we have found that forcing a 20 percent microcrystalline-collagen-suspensoid paste (partial acid salt at pH 3.2) through a Rietz Extructor unit results in the formation of uniform microparticles shown in Fig. 3.23.

When microcrystalline collagen flour is heated at atmospheric pressure up to 120°C for 30 h, the fibrils do not fuse as is shown in Fig. 3.24.   However,

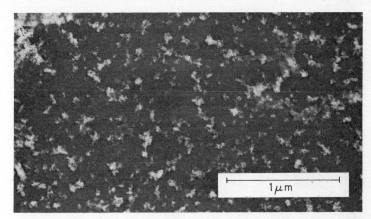

fig. 3.23 *Microcrystals in 20 percent gel of partial acid salt of bovine collagen after three passes through Rietz Extructor prior to heat treatment.*

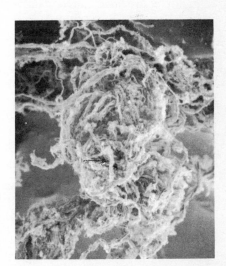

fig. 3.24 *Microcrystalline collagen flour (hemostat-adhesive form).*

if microcrystalline collagen is heated at 150 to 160°C with nominal pressure (1,000 lb/in^2), it begins to flow and fusion of the fibrils takes place (Fig. 3.25).

When the partial HCl acid salt form is subjected to an intense preheating treatment (120°C for 30 h), the morphology of the fibrils changes due to intense hydrogen bonding.   It requires considerably more mechanical energy to produce stable suspensoids from heat-pretreated microcrystalline collagen as contrasted with non-heat-pretreated microcrystalline collagen.   (See Chap. 3, Procedures.)

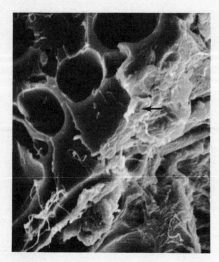

fig. 3.25   Fused microcrystalline collagen flour with loss of fibrillar morphology on fusion.

Figure 3.26 shows microcrystals in a 0.1 percent suspension of microcrystalline collagen gel (diluted after making the suspensoid at 3 percent solids in a Waring Blendor for 60 min at 25°C) prepared from a heat-pretreated (120°C for 30 h) flour.   This diluted sample had not been dialyzed to remove soluble salt prior to sampling for electron microscopy.   In contrast with this, Fig. 3.27 is an electron micrograph of microcrystals recovered from the filtrate of a 0.1 percent suspension of heat-pretreated (120°C for 30 h) microcrystalline collagen (diluted after making the suspensoid at 5 percent solids for 60 min in an Osterizer at 25°C).   Clearly, the heat pretreatment leads to novel and discrete unit microcrystal particles.   This diluted sample was dialyzed prior to sampling for electron microscopy, and a complete absence of the

1 μm

fig. 3.26   Bovine collagen microcrystal particles in suspensoid prepared from heat-pretreated (120°C for 30 h) microcrystalline collagen flour.

fig. 3.27 *Bovine collagen microcrystals after heat treatment (120°C for 30 h) and 60 min at 5 percent solids in an Osterizer (supernatant particles of solution diluted to 1 percent solids).*

salt crystals in Fig. 3.26 is noted.   Both 0.1 percent dilute suspensions were filtered through ultrafine sintered-glass filters, and the sampling was from the top of the vials containing the respective filtrates after they had been allowed to stand at least 16 h.

Figure 3.28 provides a clear picture of how the viscosity of microcrystalline collagen peaks in the 3.0- to 3.4-pH range, as well as other interrelated physical properties.   Figure 3.29 is a photograph of a 1 percent aqueous microcrystalline collagen suspensoid.   Despite the fact that this gel contains 99 percent water, it is a solid pseudoplastic, spreadable gel.

Figure 3.30 shows the viscosity of a 1 percent microcrystalline collagen gel as a function of the composition of the partial HCl acid salt, which clearly shows that a maximum viscosity level is attained within closely controlled limits of pH (corresponding to gel equilibrium pH's at 0.5 percent solids of 3.2 ± 0.2).   Figure 3.31 provides a calibration curve which interrelates the optimum initial HCl normality and concentration for increasing concentrations of microcrystalline collagen in the aqueous suspensoids.

Infrared spectra of microcrystalline collagen before and after deuteration are given in Figs. 3.32 and 3.33, respectively.   A significant change in the IR spectrum is found after deuteration, but no attempt at interpretation of these data is being made inasmuch as extensive related studies are continuing, the data from which will provide a sounder basis for interpretation.

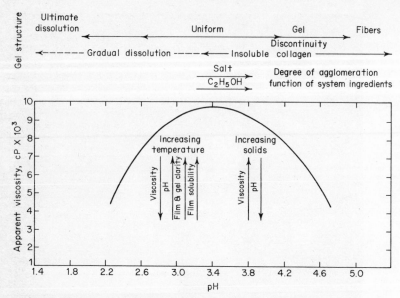

fig. 3.28 *Variations in collagen gel properties in the pH range of 2.0 to 5.0.*

fig. 3.29 *Microcrystalline collagen, 1 percent aqueous gel.*

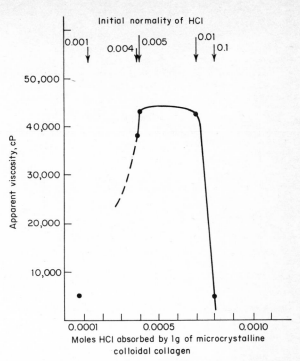

fig. 3.30 *Apparent viscosities of 1 percent microcrystalline collagen gels as the function of chemically absorbed HCl.*

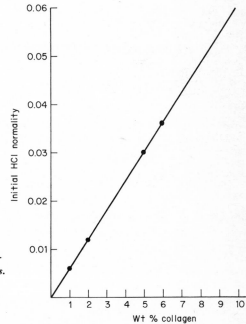

fig. 3.31 *Optimum HCl normality for microcrystalline-collagen-gel preparations.*

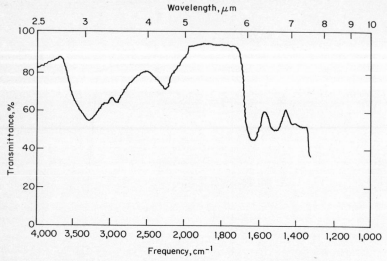

fig. 3.32 *Infrared spectrum of microcrystalline collagen before deuteration.*

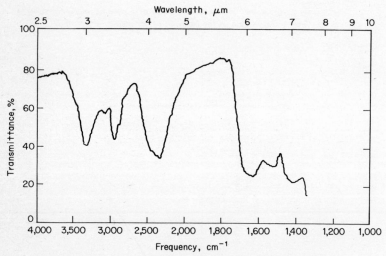

fig. 3.33 *Infrared spectrum of microcrystalline collagen after deuteration.*

APPLICATIONS

A pure-collagen product possessing a unique fibrous morphology and the capability of producing highly viscous aqueous suspensoids in water at concentrations as low as 0.5 percent would be expected to have a very wide range of uses. Experience in the development of applications of microcrystalline collagen has borne this expectation out as the material presented in this chapter reveals. Because the product is derived from a naturally occurring protein source and no toxic ingredient need be used in the preparation of the product, the product presents no problem of toxicity.

In Surgery

Microcrystallne collagen in the form of a fluffy white flour (Fig. 3.24) or in the form of a nonwoven web or sheet (Fig. 3.34) has been extensively demonstrated, both in animals and in humans, to possess unique properties

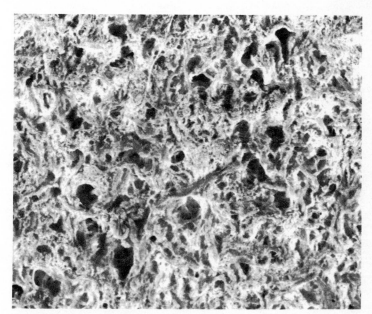

fig. 3.34   *Microcrystalline collagen flour in web or sheet form—alternate hemostat-adhesive dosage form.*

as a novel combination hemostat and tissue adhesive (54). An increasing number of publications are reporting on its safety and efficacy (57–64). All results to date project it as an important adjunct for surgery.

When microcrystalline collagen hemostat is placed *dry* on a severed biological surface of an animal and held in place with compression for 30 to 90 s (depending on the severity of the bleeding), hemostasis is produced. Figure 3.35 illustrates the microcrystalline-flour matrix before and after exposure to human blood. The blood fuses the fibrillated fibers at their interstices, which locks them into a "fishnet" matrix. As hemostasis is effected within 30 to 90 s a strong tissue-adhesive mix forms.

On the other hand, it has been clearly demonstrated that the efficacy of the microcrystalline collagen hemostat is reduced or substantially lost if the open fibrin-like fibrillar structure of Fig. 3.35 is not produced. For example, a product in which the fibrils are interlocked tightly as shown in Fig. 3.36 is not efficacious as a hemostat adhesive.

Of even greater importance is that the excess microcrystalline collagen on the wound may be removed by means of forceps without impairing the effectiveness of the hemostasis. In neurosurgical and other internal surgical procedures where the excess blood at a site must be removed vigorously by means of a vacuum tube or *sucker*, hemostasis remains intact despite probing about the site of the sealed wound.

It is most important that the agent be applied to the bleeding surface dry and held in place (at least 60 to 180 s) with compression until hemostasis results. Soaking of the agent with saline or any other liquid prior to use as above will greatly reduce its hemostatic efficacy.

Inasmuch as microcrystalline collagen is the purest form of native bovine

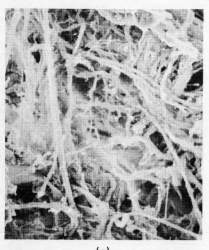

(a)   (b)

fig. 3.35   (a) *Microcrystalline collagen hemostat control; (b) microcrystalline collagen hemostat soaked with blood.*

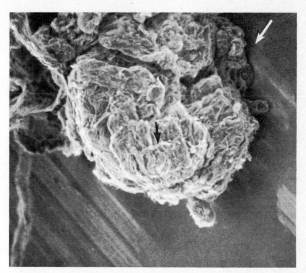

fig. 3.36   *Unopened bundles of microcrystalline collagen fibrils—not efficacious as a hemostat.*

collagen known to man, it exhibits favorable tissue compatibility.   When left inside the body subsequent to an operation, or internal surgery, experience to date has shown rapid bioassimilability (from 4 weeks to about 90 days depending on the site, blood supply, and amount required) without any unfavorable immunological response.   It is desirable to use microcrystalline collagen hemostat at sites where there is access to an adequate supply of blood so that complete bioassimilability may be effected in the shortest possible time.   There now is a growing body of literature covering the experimental and clinical experience with microcrystalline collagen hemostat adhesive. The early pioneering investigations of its application in surgery were conducted by Hait, Battista, Stark, and McCord (57); Hait (58); Lorenzetti, Fortenberry, Busby, and Uberman (59); Rucker, Kettrey, and Zeleznick (60); Hait, Robb, Baxter, Borgmann, and Tippett (61); Wilkinson, Tenery, and Zufi (62); McDonald, Kasten, Britton, Smith, Bogle, Borgmann, and Robb (63); and Lorenzetti, Fortenberry, and Busby (64).

Wound and Burn Dressings

Suspensoids (either aqueous or aqueous–organic-solvent mixtures) may be freeze-dried to produce novel skin-like wound dressings and burn bandages (48,51).

Microcrystalline collagen gels that are freeze-dried from aqueous suspensoids give products with widely varying properties depending on the solids content of the gel.   Absorbent wound pads are best prepared from 0.5 to 1.0 percent solids suspensoids.   Interestingly, they give rise to a nonfibrous

fig. 3.37   Freeze-dried aqueous microcrystalline collagen gel—note smooth film surface, which is similar to that of heat fusion (Fig. 3.25).

fused or film-surface structure as shown in Fig. 3.37.   Such a product exhibits little or no hemostat-adhesive properties on a bleeding surface.   Freeze-dried dressings prepared from 100 percent aqueous suspensoids possess a high degree of water stability or integrity.   On the other hand, if suspensoids comprising an alcohol-water medium are freeze-dried, the resulting product possesses poor integrity when immersed in water or wetted out.   Figure 3.38 explains this reduced integrity in water inasmuch as the morphology is much more open than for Fig. 3.37.

fig. 3.38   Freeze-dried aqueous-alcohol microcrystalline collagen gel.

These spongelike mats still contain all of the very small amounts of acid used in making the colloidal collagen; no neutralization whatsoever is made nor is it necessary in the case of the aqueous precursor forms. Small amounts of typical collagen cross-linking agents may be added to the suspensoid at the beginning of the attrition step.

Products made in this manner exhibit water absorption properties at least 3 times greater than surgical cotton, imbibing at least 50 times their own weight of water. For surgical purposes and as wound dressings, the products are superior to surgical cotton because they are lint-free. Their tensile strength, especially the wet tensile strength, may be improved by incorporating in the gel prior to freeze drying other fibers such as unswollen collagen hide fibers, cotton, rayon, nylon, polyesters, wool, carded freeze-dried collagen fibers, etc. The proportion of added fibers may be up to 25 percent or more based upon the weight of the microcrystalline collagen in the gel.

Typical cross-linking agents which are satisfactory include the various formaldehyde-based cross-linking agents such as, for example, urea-formaldehyde precondensate and melamine-formaldehyde precondensate, formaldehyde, glyoxal, acetaldehyde, glutaraldehyde, potassium alum, chrome alum, iron alum, basic aluminum acetate, cadmium acetate, copper nitrate, barium hydroxide, water-soluble diisocyanates, etc. The specific cross-linking agent which is utilized will be dependent upon the end use of the products although we have favored the use of glutaraldehyde and potassium aluminum sulfate. Interestingly, products cross-linked with potassium alum resisted discoloration on being heat-treated at temperatures of 150 to 200°C more than any others. Obviously, the cross-linking reactions may be accelerated by moderate heating prior to freeze drying, and this moderate heating is also advantageous where the higher concentration of microcrystalline collagen is used in that the viscosity of the dispersion may be lowered to some extent. In no instance should temperatures greater than about 90°C be employed. For medical and surgical uses the innocuous cross-linking agents such as alums are preferred.

Interesting phenomena have been observed with respect to the shrinkage behavior of these freeze-dried microcrystalline collagen mats. Mats were made from 0.75 percent gels of microcrystalline colloidal collagen, and there was included in the different mats a variety of cross-linking agents. In each instance 0.001 mol of the cross-linking agent was added per 100 g of gel. Mats containing potassium alum, melamine-formaldehyde condensate, basic aluminum acetate, cadmium acetate, chrome alum, copper nitrate, and barium hydroxide were prepared and compared with a similar untreated mat.

Measurements were made of the shrinkage of the mats on heating. Specimens were heated in an oven from 25 to 200°C by 10 or 25°C intervals, the specimens being retained in the oven for 1 h at each temperature. All of the specimens showed a shrinkage not lower than about 2.5 percent up to 100°C and 5 percent up to about 140°C. The control mat and the mat containing melamine-formaldehyde showed an increasing shrinkage which amounted to about 10 percent at 175°C and rose to 35 percent at 200°C. The mat containing copper nitrate had a 10 percent shrinkage at about 160°C and a 26 percent shrinkage at 200°C. The cadmium acetate–containing mat had a shrinkage of about 10 percent at 180°C and about 20 percent at 200°C. The potassium alum and the aluminum acetate–containing mats had a shrinkage of about 5 percent at 150°C and 10 percent at 200°C. The chrome alum mat had a shrinkage of 5 percent at 170°C and 9 percent at 200°C. The barium hydroxide–containing mat had a 5 percent shrinkage at 180°C and a 10 percent shrinkage at 200°C.

Heating of the mats also showed that no visible deterioration or change of color was noticeable at temperatures up to 100°C. Most of the mats began to exhibit a slight discoloration at 120°C, and the discoloration increased as the temperature increased. However, the mat prepared from the gel-containing potassium alum showed no visible color change at 200°C and remained white in color.

Should higher-strength mats be required, gels having higher concentrations of the microcrystalline collagen should be used because the strength of the products varies directly with the concentration of the gels. The water absorption of the products, however, varies inversely with the concentration of the gels; and, accordingly, for specific applications, it is necessary to take both properties into consideration in the preparation of the original gel.

Other projected medical uses of microcrystalline collagen in various structural forms are illustrated in Fig. 3.39.

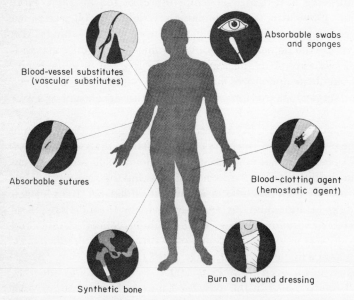

Absorbable swabs and sponges

Blood-vessel substitutes (vascular substitutes)

Absorbable sutures

Blood-clotting agent (hemostatic agent)

Synthetic bone

Burn and wound dressing

fig. 3.39 *Microcrystalline collagen, bioengineerable natural material for medical and surgical uses.*

Microcrystalline Collagen–Apatite Synthetic Bone Structures

The principal organic constituent of cartilage and bone tissue is collagen. The principal inorganic component in cartilage and bone tissue is calcium phosphate complexes or compounds such as, for example, hydroxyapatite. The chief inorganic constituents are calcium, magnesium, phosphate radicals, carbonate radicals, the fluoride radical, and water—the compounds being of varying compositions generally belonging to the apatite group. Other inorganic ions are generally present in trace amounts, and other organic matter is also present. The other elements found in minor or trace amounts in bone tissue are aluminum, barium, boron, chlorine, copper, iron, lead, manganese, potassium, sodium, strontium, and tin, while arsenic, bismuth, lithium, molybdenum, nickel, selenium, silicon, silver, and zinc have been detected spectrographically. In general, the differences in hardness and rigidity between cartilage and bone tissue are due to the differences in composition such as differences in the ratios of collagen to the inorganic calcium phosphate compounds and the presence of other radicals as well as the morphological structure.

Thus far, cartilage and bone tissue have not been formed or duplicated synthetically. In bone surgery, a variety of materials has been used including bone, bone derivatives, and synthetic substitutes. Bone from which certain constituents such as minerals, proteins, lipids, and water have been removed is generally classed as bone derivative. Synthetic substitutes include metals, certain synthetic polymers, calcium sulfate and hydroxyapatite, and porous ceramics.

The chemistry of sparingly soluble phosphate salts or specifically the system $Ca—P_2O_5—2H_2O$ and the precise chemistry and structure of the calcium phosphate compounds occurring in natural cartilage and bone tissue are extremely complex. Accordingly, the term *calcium phosphate* as used herein includes dicalcium phosphate, tricalcium phosphate, octacalcium phosphate, hydroxyapatite, carbonate-apatite, chlorapatite, fluorapatite, and mixtures thereof.

Two major types of natural bone are well recognized: very hard, high tensile strength bone known as *cortical bone* and the more porous spongelike form called *cancellous bone*. Figures 3.40 and 3.41 are electron-scan micrographs of natural bovine cortical bone and natural bovine cancellous bone, respectively.

Bonelike structures may be formed from a homogeneous mixture of a water-insoluble, ionizable microcrystalline salt of collagen, calcium phosphate, and water (47).

In forming synthetic cartilage-like and bonelike structures, mesomorphous calcium phosphate is mixed with an aqueous gel of the water-insoluble, ionizable microcrystalline collagen salt. Since bone tissues contain some citric acid and phosphate salts, the microcrystalline collagen is preferably prepared

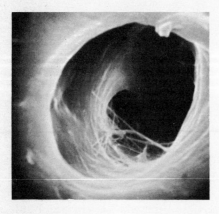

fig. 3.40 Bovine cortical bone.

by treatment of the collagen source material with citric acid or phosphoric acid or an acid phosphate salt so as to avoid the presence of what may be termed foreign ions in the product. The calcium phosphate may be formed by mixing solutions of a soluble phosphate, such as, for example, sodium phosphate, and a soluble calcium salt, such as calcium acetate, so as to provide the desired molar ratio of calcium to phosphate, preferably about 1.6 : 1.

In the biological formation of bone tissue and particularly the hard tissue of teeth, a carbonate-apatite compound is produced which apparently accounts for the extreme hardness of such tissue. In the production of the structures according to this invention, desired amounts of the various anions may be incorporated in the gel mixtures, preferably in the preparation of the calcium phosphate, to provide relative ratios approximating those of natural bone tissue. In the addition of the carbonate anion and the fluoride anion, for example, about 1 to 10 molar percent, preferably 2 to 5 percent, of the phosphate anion may be replaced by the carbonate anion and about 0.05 to 2 molar percent, preferably 0.1 to 0.6 percent, of the phosphate ion

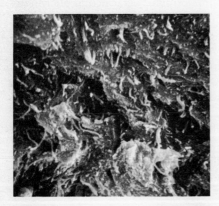

fig. 3.41 Bovine cancellous bone.

may be replaced by the fluoride anion. Where it is desired to incorporate other ions such as the carbonate and fluoride anions, solutions of salts such as sodium carbonate and sodium fluoride are added during the preparation of the calcium phosphate. Preferably these solutions are added to the soluble phosphate solution so that upon mixing with the solution of the calcium salt the salts containing the other ions will coprecipitate with the calcium phosphate.

The solutions of the soluble salts are mixed under vigorous agitation. The resulting slurry of the insoluble salts will generally have an alkaline pH of from 10.5 to about 12. The mesomorphous calcium phosphate is separated by filtration and washed thoroughly with water so as to remove soluble salts. The recovered wet salt is slurried in water, and just prior to the addition of the slurry to an aqueous gel of the microcrystalline collagen salt, the pH of the slurry is adjusted to a pH of between about 3 to 5 by the addition of acetic acid.

Where the ratio of the calcium phosphate to collagen is within the upper portion of the range, the mixture after drying forms a white bonelike structure. Where the ratio of the calcium phosphate to collagen is in the lower portion of the range, the mixture after standing overnight at a temperature of about 5°C separates with a nongelatinous mass at the top which, when recovered from the lower water layer and allowed to dry, forms an amber-colored cartilage-like structure.

Assuming a fixed composition of the collagen gel–calcium phosphate mixture, for example, equal parts by weight of microcrystalline collagen and calcium phosphate, the porosity of the dried product formed therefrom may be altered by varying the amount of dispersed air or gas in the gel mixture. A gel mixture containing substantially no dispersed gas when oven-dried produces a very hard, dense product resembling cortical bone. An aerated gel mixture containing 40 vol percent dispersed air or gas, when oven-dried, forms a product having a very porous interior structure, but the surface portions are of an appreciably lower porosity and are hard and somewhat horn-like. On the other hand, the same gel mixture containing 40 vol percent dispersed air or gas, when freeze-dried, forms a product of substantially uniform porosity; the surface portions are of approximately the same structure as the interior, and the product resembles cancellous bone. Porosity may also be controlled by the concentration of solids in the gel mixture. In general, the higher the microcrystalline collagen content of the gel mixture, the greater the compactness of the product. Gel mixtures containing 20 percent or more microcrystalline collagen form hard, compact products, whereas those containing a maximum of 5 percent microcrystalline collagen form very porous structures. Structures formed by freeze drying exhibit the greatest porosity; those formed by air drying a lower porosity; and those formed by oven drying the lowest porosity, as would be expected.

The solids content of the microcrystalline-collagen-salt dispersion or soli-quoid-type gel may be varied depending upon the nature of the desired product. Stable nonelastic-type gels may be formed with as low as 0.5 wt percent of the microcrystalline collagen salt. For the cartilage- and bone-type products, the gels may contain up to about 30 percent or more microcrystalline collagen salt.

Aqueous gels prepared from the water-insoluble, ionizable microcrystalline collagen salts exhibit a pH from about 2.6 to about 3.8, the gels exhibiting optimum properties having a pH between about 3.0 and 3.4. Fortuitously, mesomorphous calcium phosphate is transformed to the mesocrystalline state at room temperature at pH's between about 3.5 and 5.5. Accordingly, in the mixtures of the microcrystalline-collagen-salt dispersion and the calcium phosphate slurry, the pH may be such that about optimum gel conditions may be maintained for the alteration of the crystal structure of the calcium phosphate. Due to the complexity of the nucleation process which is common in the case of sparingly soluble salts, the calcium ions and the phosphate ions interact with the colloidal collagen particles. This interaction involves a solution transport mechanism of the ions to the nucleation sites furnished by the surfaces of the colloidal collagen particles.

In order to form water-insoluble and structurally stable products, the microcrystalline colloidal collagen must be in a form such as will allow an interaction with the calcium phosphate inorganic constituents. The colloidal microcrystalline collagen dispersion or gel must contain the calcium and phosphate ions in such a state that they can interact to alter the crystalline structure of the inorganic constituents. The product, in other words, must have a very intimate and uniform distribution of the various constituents.

The size and shape of the calcium phosphate crystals may be modified by the proportion of carbonate anions present. In general, hydroxyapatite crystallizes in needle form and the presence of small proportions of carbonate ions, such as up to about 25 percent carbonate anions based on the weight of the hydroxyapatite, tends to reduce the length of the crystals. As noted above, the gel of the collagen salt has a pH of 3.0 to 3.4, and the pH of the gel may be adjusted, that is, raised to at least 3.5 to provide conditions favorable for the alteration of the structure of calcium phosphate.

The mixtures are dried to a moisture or water content from about 5 to about 25 percent. Where porous structures are desired, drying is preferably effected by a freeze-drying process. After the moisture content has been reduced to a value within this range, the structures are heated, for example, in an oven in an inert atmosphere, to a temperature between about 100 to about 150°C for 2 to 10 h. This heating step substantially increases the stability of the structural characteristics of the products in the presence of aqueous liquids.

The following examples describe the experimental procedures for preparing two typical forms of "synthetic bone" from microcrystalline collagen (47).

EXAMPLE 3.3   A calcium phosphate slurry is formed by first dissolving 4.78 parts of trisodium phosphate in 100 parts of deionized water. The trisodium phosphate solution is then mixed with a solution containing 17.6 parts of calcium acetate in 100 parts of deionized water with agitation. The resulting slurry will have a pH of 11.3. The salt is recovered, washed to remove soluble salts, and slurried in deionized water, and the pH is subsequently reduced to about pH 3.2 by the addition of acetic acid. The microcrystalline collagen is prepared by treating 1 part by weight of ground vacuum freeze-dried bovine collagen with 100 parts of an aqueous solution of hydrochloric acid having a pH of about 2.4. The treated bovine collagen is transferred to a Waring Blendor where it is attrited for about 25 minutes at a temperature which is maintained below 25°C. The resulting gel will have a pH of about 3.6.

A mixture of 250 parts of the calcium phosphate slurry and 250 parts of the microcrystalline collagen gel is made with agitation. The resulting mixture is placed in a refrigerator, cooled to about 5°C, and maintained in the refrigerator overnight. A semisolid gel-like material collects at the surface of the liquid, which is separated from the liquid and dried in an oven. This gives an amber-colored product that has the appearance and feel of cartilage.

Another sample prepared in the same manner, when freeze-dried for 12 h to a water content of about 10 percent (-40 to -50°C, vacuum 5 /μm, heating cycle not exceeding 30°C with condensation of sublimed water at 60°C), leads to a porous structure. After the freeze-dried product is obtained, it is placed in an oven and heated to a temperature of about 103°C for 3 h.

Both products prepared as above retain their structure upon continued immersion in water. The initial cartilage-like product becomes rather soft and very pliable, while the heat-treated product remains firm but pliable. By analysis, the products will contain 2.54 percent calcium, 1.38 percent phosphorus, and 12.96 percent nitrogen.

EXAMPLE 3.4   Calcium phosphate is formed by mixing trisodium phosphate and calcium acetate solutions as described in the preceding example. After thoroughly washing the precipitated salt, it is slurried in deionized water to form a slurry containing approximately 25 wt percent calcium phosphate.

Microcrystalline collagen salt of citric acid is formed by slurrying a ground vacuum freeze-dried bovine collagen in water, and the slurry is then centrifuged to provide a slurry containing approximately 31 percent of the dispersed ground collagen. A mixture of equal parts of isopropanol and water is then added to reduce the solids content of the slurry to approximately 10 percent. A mixture of 2,390 g of this slurry and 814 g of isopropanol is made. Eight hundred ml of a 1 N solution of citric acid in isopropanol are then added to the slurry during continuous agitation. The pH of the liquid is approximately 2.9. The slurry is then centrifuged to reduce the solids content to about 31 percent. Additional amounts of isopropanol are added, and the slurry is then subjected to centrifuging to remove as much liquid as possible. The residue is air-dried.

The air-dried product is mixed with deionized water and subjected to attrition in a blender-type mixer at a temperature below 25°C to form a suspensoid containing 6 percent of the disintegrated microcrystalline collagen salt.

The pH of the slurry of calcium phosphate is next adjusted to a pH of 3 by the addition of acetic acid. To 300 parts of this suspensoid there is added a sufficient amount of the calcium phosphate slurry to provide 18 parts of calcium phosphate. The mixture is subjected to vigorous agitation, then deaerated and poured into a pan to a depth of about $3/_4$ in, and allowed to dry in the air for about 1 week. The suspensoid formed has an intense white appearance and, when dried, exhibits appreciable shrinkage, but the product is stiff and hard. When the product is cut and the cut surface is examined under the microscope, it exhibits a fine, porous spongelike structure. By analysis, the product contains 21.46 percent calcium, 9.73 percent phosphorus, and 7.88 percent nitrogen. A portion of the product is heated in an oven to a temperature of about 103°C for 2 h. Samples of the original product and the heat-treated product retain their structure on continued immersion in water.

Preliminary tests in animals with synthetic-bone materials made from microcrystalline collagen by these and similar procedures have shown excellent tissue compatibility and no significant interference with osteogenesis of new bone according to more than one protocol or test method. It is still premature to claim clinical orthopedic value for these types of synthetic cancellous and cortical bone materials.

Figure 3.42 illustrates cortical forms of synthetic bone, and an electron micrograph of this type of dense bone material is shown in Fig. 3.43. Similarly, Fig. 3.44 illustrates synthetic bone of the more porous structure, and Fig. 3.45 is an electron-scan micrograph showing its open, porous morphology.

This much can be said—orthopedic surgery awaits the day when a biocompatible and controlled bioassimilable prosthesis will be available for clinical use. On general principles nonassimilable prostheses such as metals and plastics leave much to be desired inasmuch as they may have to be implanted in a patient for life. This is not of great concern in the elderly, but it may be a problem when patients are children or young adults and the inert prosthesis may have to remain in the body for decades rather than for relatively few years.

Colloidal Organic-Solvent–Microcrystalline Collagen Dispersions

Stable colloidal dispersions may be made of water-insoluble microcrystalline collagen in dispersing media consisting of (1) dimethyl sulfoxide, (2) mixtures of water and dimethyl sulfoxide, (3) dimethyl sulfoxide–free mixtures of water

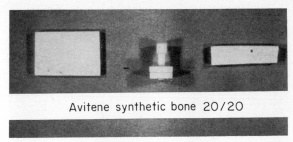

Avitene synthetic bone 20/20

fig. 3.42   *Synthetic (Avibone) microcrystalline collagen—cortical-type bone.*

and up to 65 wt percent of a water-miscible organic solvent, or (4) mixtures of water, dimethyl sulfoxide, and up to 75 wt percent of at least one other water-miscible organic solvent (65).   The colloidal dispersions are formed by treating collagen with a dilute acid and attriting until at least 10 wt percent of the collagen has been reduced to a colloidal submicron size and dispersing such attrited material in a dispersing media or alternatively by treating the collagen with the dispersing media containing the acid and attriting in the presence of the dispersing media.

Water-miscible organic solvents, on being added to aqueous dispersions of this form of collagen, do not precipitate as would be expected but first actually thin out the dispersion somewhat.   As the amount of water-miscible solvent added is increased, the dispersion finally reaches a concentration where the collagen precipitates.   This varies with the solvent and for any solvent with the concentration of the collagen in the system.   In general, the useful range for solvent concentration is up to about 65 parts by weight.

fig. 3.43   *Electron-scan micrograph of synthetic microcrystalline collagen—cortical-type bone.*

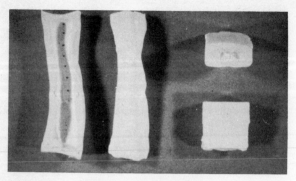

Avitene synthetic bone 10/10

fig. 3.44   *Synthetic microcrystalline collagen—cancellous-type bone.*

While even very small amounts of additional solvent serve the purpose of preventing bacterial action, the other advantages of the use of added solvent are not obtainable below about 20 wt percent of solvent. In general, we prefer to operate with 25 to 35 percent organic solvent in order to prevent possible local concentration of substantially above 65 percent, which can cause precipitation of the collagen.

Dimethyl sulfoxide is an exception to the general rule in that it will not cause precipitation even when used alone; it acts more like water than like the ordinary water-miscible organic solvents. Moreover, when it is used in conjunction with water, it permits the incorporation of more than an additional 50 percent of organic solvent. Thus a 50 percent isopropanol/50

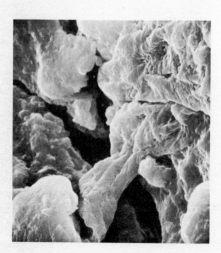

fig. 3.45   *Electron-scan micrograph of synthetic microcrystalline collagen—cancellous-type bone.*

percent water medium is on the edge of precipitation. On the contrary, a 50 percent isopropanol/25 percent dimethyl sulfoxide/25 percent water mixture gives a clear homogeneous dispersion; in fact, the percentage of isopropanol can be substantially increased, in this instance up to about 75 percent. Dimethyl sulfoxide alone is a good dispersing medium. However, when water-miscible organic solvents such as isopropanol are added to dimethyl sulfoxide in the absence of water, then the addition of as little as about 25 percent isopropanol to the dimethyl sulfoxide will cause precipitation of the microcrystalline collagen.

Not only can the water-miscible solvent be added to an already existing dispersion, but the product can be made by direct treatment of hide substance with a mixture of dilute aqueous acid and the water-miscible solvent at the optimum concentration of acid, water, and solvent in conjunction with the required mechanical disintegration treatment.

We have found that the phenomenon discussed above is general for substantially all water-miscible solvents with the above-mentioned exception of dimethyl sulfoxide. We have investigated aliphatic alcohols, such as methanol, ethanol, isopropanol, n-propanol; cyclic alcohols such as tetrahydrofurfuryl alcohol and furfuryl alcohol; water-soluble ethers such as dioxane and tetrahydrofurane; and ketones such as acetone and methyl ethyl ketone.

As indicated above, the dispersions of this invention can be made either by adding the solvent after an aqueous gel of microcrystalline colloidal collagen has been made or by adding the solvent directly to the starting mixture of hide and dilute acid. The second of these methods invariably yields a dispersion of lower viscosity. This difference in viscosity becomes more marked as the concentration increases and/or the temperature increases up to about 80°C.

The use of increases in temperature, above ambient temperatures up to about 80°C, to control viscosity temporarily is an unexpected advantage of this invention.

One of the major advantages of this invention is that these solvent-containing gels are completely free from bacterial attack and have a stable viscosity, not only over extended periods of time but over all likely storage temperatures, even in the summer. The other major advantage of using organic-solvent diluents is that fibers can be dry-spun and films can be dry-cast because the organic solvents generally have lower boiling points and lower heats of evaporation than water and thus the drying time is shortened. This is particularly true where articles are to be dip-coated.

Other advantages are (1) the solvent gels are more easily deaerated than the aqueous gels and (2) the solvents can be used as carriers for additives, such as cross-linking agents, thermal stabilizers, dyes, photosensitive compounds, and the like, some of which tend to be water-insoluble.

Tensile properties of films made with the four formulations are summarized in Table 3.15. Tensile strengths ranged from about 10,000 lb/in^2 for films without formaldehyde to about 11,000 lb/in^2 for those with formaldehyde. The 10 percent difference is attributable to cross-linking of collagen by formaldehyde. There were no differences in tensile elongation at break; they were all about 5 percent.

There is no significant difference between the tensile properties of films prepared from dispersions with or without isopropanol. However, the isopropanol films did possess greater clarity and they appeared to be more uniform. Two processing advantages of using isopropanol dispersions in this

table 3.15   Mechanical Properties of Hand-cast Films Prepared from Collagen Dispersions (65)

Dispersion formulation	Tensile strength, lb/in^2	Elastic modulus, lb/in$^2 \times 10^{-3}$	Tensile elongation at break, %
1% collagen gel in 0.01	11,100	3.72	0.0
N HCl	8,400	4.03	2.8
	2,800	3.74	0.83
	12,900	4.22	6.0
	13,000	4.02	7.9
	6,600	4.06	1.9
	9,900	3.90	3.6
	8,700	4.03	3.0
Averages	9,600	3.97	4.39
1% collagen gel in 0.01	13,200	3.95	7.1
N HCl with 2% CH$_2$O	9,700	3.85	3.7
	10,300	3.70	4.3
	10,000	3.55	4.5
	11,100	3.77	5.8
Averages	10,900	3.76	5.11
1% collagen gel in 50%	7,800	3.17	4.3
0.01 N HCl and 50%	9,400	3.42	5.3
isopropyl alcohol	9,600	3.82	4.6
	10,500	3.91	6.3
	10,000	3.76	5.3
Averages	9,470	3.62	5.11
1% collagen gel in 50%	8,700	3.75	3.6
0.01 N HCl and 50%	10,900	4.16	5.7
isopropyl alcohol with 2%	12,700	4.26	7.3
CH$_2$O	10,600	4.64	3.6
	11,400	4.47	5.0
Averages	10,900	4.26	5.04

application are that their viscosity is lower and they support higher loadings, which makes possible the casting of thicker films by a single step.

Microcrystalline colloidal collagen is useful in the preparation of meat products such as sausage and meat loaf. The gel may be used to control the texture, juiciness, and organoleptic characteristics and is particularly adapted to effect a uniform distribution of seasoning ingredients. The seasoning ingredients are dissolved and/or dispersed in the gel, and the gel is then used in preparing the meat emulsion for stuffing into casings or in preparing the meat loaf.

Pharmaceuticals (49,53)

Microcrystalline colloidal collagen is highly satisfactory in a wide range of pharmaceutical and cosmetic compositions or preparations. It is particularly advantageous in liquid-, paste-, and cream-type compositions because of the ability to form gels at low concentrations and the ability to maintain insoluble ingredients in a stable dispersed state. It has a high affinity or sorptive power for oleaginous substances and will function as a means for dispersing the oleaginous substances in aqueous liquids. In lotion-, emulsion-, cream-, and ointment-type compositions that conventionally include oily, fatty, or waxy (oleaginous) ingredients, the stability of the composition may be improved by the addition of a small proportion of the colloidal collagen. Microcrystalline colloidal collagen may be utilized to replace part or all of the oleaginous ingredient. The ability to reduce or eliminate oleaginous ingredients in these types of compositions reduces the greasiness of the products. Accordingly, after the composition is spread over the skin of the user and allowed to dry, the coatings have a lower soiling tendency and the coatings are not readily transferred by contact with clothing, bandages, etc.

The new form of collagen has excellent compressibility characteristics and, when formed under compression, provides a coherent structure and, accordingly, finds value in the manufacture of tablets and compressed shapes. It may be used to replace in whole or in part conventional substances found in tablets such as starch, sugar, and binding agents. Also, in powder composition, this form of collagen may replace, in whole or in part, opacifying agents (clay, magnesia, zinc oxide, etc.), slip materials (talc, metal stearates, etc.), adherent materials (clay, stearates, etc.), and absorbents (chalk, kaolin, etc.).

Cosmetics (53,66)

A wide selection of cosmetics ranging from gels, creams, and lotions to lipstick and shaving cream depend for their aesthetic and functional sales success on properties contributed to them by colloidal or submicron particles. Such colloidal particles control, in part, the stability, smoothness, and viscosity of hundreds of cosmetics. Inasmuch as microcrystalline collagen gels comprise suspensions of submicron particles, they present themselves as novel and unique vehicles for a range of cosmetic applications.

Microcrystalline collagen has been used successfully in the laboratory to formulate a wide range of cosmetic products where its thixotropic and pseudoplastic rheology, its nonpilling and film-forming properties, and its unique water-holding capacity are advantageous. The compatibility of this form of collagen in the presence of high proportions of organic solvents including isopropanol and ethyl alcohol adds to its potential in cosmetics. For example, microcrystalline collagen has been laboratory tested in the following typical cosmetic formulations: face powders, creams, sunscreen lotions (nongreasy, water-dispersible types), cleansing lotions, deodorants, skin-masking preparations, toothpastes, aerosol products such as shaving creams, greaseless ointments, and stable calamine preparations. It proffers opportunities for engineering new product concepts in cosmetics generally.

Microcrystalline collagen in the form of aerosol hair preparations is advantageous because it is bioassimilable and, unlike other film-forming polymers used in such preparations, does not accumulate in the body as a foreign material.

In shaving creams, soaps, bubble bath compositions, etc., this form of collagen has utility because of its ability to strengthen and stabilize the lather or foam. It appears to function to strengthen the gas-enclosing walls and prevent them from readily breaking down.

Foods (53)

Microcrystalline colloidal collagen is particularly useful in a wide range of food products because of its properties. It is edible and nutritious, is bland in both taste and odor, has very little, if any, color and is free of textural defects which could adversely affect the taste and mouth feel of the food products. Gels have a very smooth eating quality and, when present in foodstuffs, become an indistinguishable part of the product. Their viscosity-building properties are of an order of magnitude greater than gelatin (Fig. 3.46).

Small proportions of microcrystalline colloidal collagen form highly viscous thixotropic gels which allow improvements in various physical properties of foodstuffs such as eating quality, visual and tactual qualities, and visual and taste textures. Thus, for example, liquid foodstuffs such as salad dressings may be thickened and insoluble ingredients may be maintained in a relatively uniform dispersed state by incorporation of small proportions of the microcrystalline colloidal collagen. This form of collagen may replace gelatin to produce products which will have a substantially stable viscosity over a long period without a gradual increase in viscosity and setting into a gummy mass as occurs with gelatin. The product is particularly well suited as an inexpensive replacement for egg white and gelatin in various types of food

fig. 3.46 Relative viscosity differences between microcrystalline collagen (Avitene) and gelatin.

products because it produces the same effects at appreciably lower concentrations.

Industrial Uses—Coatings and Photographic Emulsions (52,67,68)

Suspensoids of microcrystalline collagen in aqueous media or in aqueous–organic-solvent media are especially useful for a host of unusual coating applications.

Such gels when applied to a structure or base material dry down to produce self-adherent continuous coatings.   In forming a coating composition for any specific purpose, the composition is formulated to meet the conditions of use of the finished product.   For example, a wide variety of cross-linking agents may be included such as potassium alum, melamine-formaldehyde precondensate, basic aluminum acetate, cadmium acetate, chrome alum, copper nitrate, barium hydroxide, etc.   These cross-linking agents improve the strength and toughness of the films and will also improve the heat resistance of the coating films with respect to both dimensional stability and resistance to discoloration.   Coatings containing potassium alum as a cross-linking agent exhibit the highest resistance to discoloration.

Such collagen coatings are useful in making grease-proof papers in thicknesses as low as 0.0003 in; they have excellent adhesion to paper and are flexible, strong, and durable.   They do not crack on handling and consequently are highly useful for such coatings.   The dispersions or gels may also be admixed with conventional coating pigments and loading agents, such as calcium carbonate, to produce coated papers.

Not only is this form of collagen useful as a coating agent for fibrous sheets, but it forms an excellent intermediate coating for the production of continuous films such as moistureproof cellophane, being exceptionally adherent to the cellophane and accepting moistureproof coatings and films such as nitrocellulose coatings. In this type of application, the microcrystalline collagen coating serves as an anchoring agent.

Numerous uses as a film-coating material have been proposed for microcrystalline collagen (52). However, one use in particular is deserving of special consideration because of its potential industrial significance.

Inasmuch as microcrystalline collagen represents such an unusually pure form of native high-molecular-weight bovine collagen, impurities such as lipids and noncollagenous constituents have been reduced to a minimum. Furthermore, the original molecular weight is retained at a maximum while partial acid salts can be formed to control pH so that clear aqueous suspensions can be obtained with a maximum level of compatibility with other salts, etc.

Very early in the development of microcrystalline collagen coatings it was envisioned that this new ultrapure form of collagen might be especially applicable for use as a vehicle for photographic emulsions applied to paper, including the production of color films and papers wherein coupler compounds are included in the photographic emulsion or dispersion and of overcoatings or protective coatings applied after the developing and processing of the negative films, transparencies, and prints.

In the preparation of conventional photographic films, plates, and papers, the photosensitive material is dispersed in gelatin and the gelatin applied to the desired base. Microcrystalline colloidal collagen suspensoids may replace a part of or all the gelatin in these photographic emulsions. Furthermore, approximately 1 part of microcrystalline colloidal collagen is equivalent to 10 parts of photographic gelatin. It is obvious from this that appreciably thinner coatings may be formed with a resulting improved definition of the photographic image. The microcrystalline colloidal collagen coatings of this invention are insoluble in water but will imbibe water to form highly viscous gels which can be combined with the conventional ingredients of photographic emulsions to produce photographic films and papers superior in properties to those made from gelatin.

Photographic emulsions conventionally contain small quantities of materials such as alum, which is a cross-linking agent for the gelatin. These materials can likewise be used in making the emulsions using microcrystalline colloidal collagen to replace all or part of the gelatin. The cross-linked material is somewhat less sensitive to water than the non-cross-linked material.

It is not essential that the microcrystalline colloidal collagen be prepared in advance of processing. The product can be prepared *in situ* in the process of making the photographic emulsion. The following gives actual experimental details (52) as to how this may be done.

Ground predried bovine corium was ether-extracted for 1 h so as to remove lipids and then dried in a vacuum oven at 35°C to remove the ether. A mixture comprising 150 ml of water, 3 g of citric acid, 15 g of the treated hide flour, and 24 g of potassium bromide was prepared by first attriting the citric acid, hide flour, and water in a Waring Blendor for 15 min, then adding the potassium bromide, and blending the mixture at high speed for 15 min to convert the hide flour to the point where a substantial portion of it had become microcrystalline colloidal collagen. The mixture was filtered through a double thickness of cheesecloth and returned to the Waring Blendor.

A solution of 30 g of silver nitrate in 50 ml of water was slowly added to the collagen-bromide dispersion in the blender at low speed for 5 min. Then 20 g of ethanol was added, and the blender was run at a high speed for 15 min for thorough mixing. The mixture was transferred to a bell jar and evacuated to remove air using a water aspirator. Several sheets of paper were coated with the dispersion by brushing. The remaining portion of the dispersion was poured into an enameled tray, gelled by chilling with ice, shredded, transferred to a bag, and washed with water at 18°C to remove the potassium bromide and the potassium nitrate. After washing, the gel was heated to about 60°C, 20 ml of ethanol were added and then mixed, and the resulting dispersion evacuated. Paper was coated with the dispersion and then air-dried.

Finished prints prepared from this photographic paper exhibited sharp definition and satisfactory contrast and brightness.

An alternative to the complete replacement of gelatin for photography is, of course, partial replacement of gelatin. A typical example of this route (52,67,68) is as follows:

A combination of 20 g of potassium bromide and 20 g of citric acid was dissolved in 90 g of water, and $7\frac{1}{2}$ g of photographic gelatin was soaked in the salt solution until softened. The mixture was then heated to 50°C to dissolve the gelatin and subsequently cooled to 30°C.

Twenty-five grams of photographic-grade silver nitrate (free of copper, mercury, and organic dirt) was dissolved in 175 g of water at 30°C, and just enough ammonium hydroxide (20 ml) was added to redissolve the silver oxide.

The silver nitrate solution was added to the potassium bromide solution at 30°C; the mixture was digested for 10 min with rapid stirring and then poured into a solution of $7\frac{1}{2}$ g of gelatin in 30 g of water. The temperature of the mixture was raised to 50°C and digested for 10 min. It was then poured into an enameled tray surrounded by ice to set the emulsion. The gel was shredded, placed in a clean cloth bag, and placed in cold running water (temperature 15°C) for at least 1 h. The washed gel was melted at 40°C, and 100 ml of 1.5 percent microcrystalline colloidal collagen dispersion was added at 40°C, followed by 28 ml ethanol and 1 ml of 10 percent chrome alum solution. The microcrystalline colloidal collagen had been prepared from ground predried cowhide which had been ether-extracted to remove lipids. The collagen dispersion was prepared by attriting the collagen in a citric acid solution in a Waring Blendor for 15 min.

A standard photographic gelatin formulation was prepared in a like manner but differed from the above formulation in that in lieu of the addition of the microcrystalline colloidal dispersion, a solution of 23 g of gelatin in 80 g of water was added to the washed gel.

The composition of the emulsions was as follows:

		Standard, *gelatin* *only*
Gelatin (photographic-grade), g	15	38
Microcrystalline collagen, g	1.5	
Water, g	395	375
Potassium bromide, g	20	20
Silver nitrate, g	25	25
Citric acid, g	20	20
Ethanol, ml	28	28
Chrome alum (10% sol. in water), ml	1	1
Conc. ammonia (lab. reagent), ml	≈ 20	≈ 20

It should be noted that 1.5 g of the microcrystalline colloidal collagen and 20 g of water was the equivalent of 23 g of photographic gelatin in the above formulation.

The finished emulsions were applied to heavy paper and dried to form photographic-print paper. On comparison of the papers made in accordance with this invention against the standard, it was found that the emulsion layer utilizing the microcrystalline colloidal collagen was substantially thinner than that of the control. The film was tougher and more flexible and had better abrasion resistance. Comparison of prints made by projection printing of the identical negative followed by conventional development techniques showed sharper definition and more faithful reproduction of the negative than the standard.

Table 3.16 lists some of the more significant demonstrated and/or projected uses for microcrystalline collagen. It is anticipated that many additional applications—medical, cosmetic, food, and industrial—for this new physical form of ultrapure collagen will be found as the development of its commercial uses continues.

table 3.16 Demonstrated and Projected Uses for
Microcrystalline Collagen

Use No.	Description
1	Hemostat
2	Sutures (novel route)
3	Blood vessels
4	Heart valves
5	Comfort mats
6	Bed sheets
7	Capsules
8	Dialysis membranes
9	Contact lenses
10	Paper coatings
11	Biodegradable rayons
12	Sanitary bottle closures (hybrids)
13	Cigarette filters
14	Photography
15	Paper coatings and binders
16	Anchoring agents
17	Textile sizes
18	Sausage casings (novel route)
19	Regenerated leather products
20	Edible packaging films
21	Water-dispersible films (hybrids)
22	Suspending agent in inks, paints, and pharmaceuticals
23	Foods:
	Stabilizer for mustard, ketchup, etc.
	Synthetic meats
	Ice-cream stabilizer
	Egg-white substitute
	Texturizer for bread and baked goods
24	Microencapsulation
25	Cosmetics

CHAPTER FOUR

Microcrystalline Silicates

INTRODUCTION

A variety of fibrous minerals are known which are generally called asbestos. Of these, the fibrous form of serpentine—hydrated magnesium silicate—known as chrysotile is perhaps the most abundant, as well as the most versatile, principally because of the length and strength of the individual fibers. The product is widely used in preparations where its fibrous nature combined with its fire resistance gives it a marked advantage over other materials, for example, in the making of fire-resistant textiles.

In the preparation of useful articles from the crude chrysotile, the problem of separating the mineral into usable fibers has involved considerable effort on the part of asbestos chemists. Mechanical breaking of the ore masses results in the production of shortened fibers, including a considerable percentage of powder, which has been considered of little value.

Chrysotile, as produced by nature, comprises fibers containing bundles of linear fibrils in an inorganic rocklike matrix. The individual fibrils may have diameters as small as a few hundred angstroms, while their lengths may be hundreds of microns or much longer. The fibers are generally recovered from the serpentine rock deposits by a variety of mechanical crushing and screening techniques which are designed to preserve the natural length of the fibers as much as possible. Substantially all of the effort has been in the direction of preserving maximum fiber length in order to get the maximum advantage from a physical entanglement of the long fibers or fibrils in making paperlike sheets or asbestos yarns or fabrics. Figure 4.1 is an electron micrograph of individually dispersed natural chrysotile fibrils.

The major efforts in improving dispersions of long-fibered asbestos have involved the use of surfactants to permit easier separation of the fibers into individual fibrils of maximum length, usually of the order of tens of microns (69). This approach leads to fibers and fibrils containing an organophilic surface coating which, of course, may be quite undesirable, especially where the inherent high-temperature properties of asbestos are to be utilized to maximum advantage or the presence of the surfactant coating leads to poor bonding in a structural matrix.

Microcrystalline silicates are a family of novel products produced by deliberately going in the opposite direction, namely, breaking down the long fibrils into fragments, microcrystals, of truly colloidal dimensions. An appropriate combination of chemical action and mechanical attrition produces a new product characterized by submicron particle size and colloidal properties, as demonstrated by the ability to form smooth gels in very low concen-

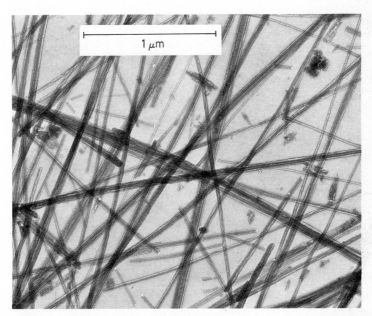

fig. 4.1 *Single chrysotile-mineral-silicate fibrils.*

Microcrystalline mineral silicate

fig. 4.2   Microcrystalline mineral silicate, 10 percent aqueous suspensoid.

trations in water and other highly polar liquids (Fig. 4.2). These smooth gels can be spread out by means of a doctor blade and will dry to adherent, cohesive, self-supporting filmlike masses without the addition of further binder. The films resemble paper.

In addition to making films and sheets, the colloidal microcrystalline silicate particles are especially effective for thickening polar compositions for many industrial uses. Conventional fibrous chrysotile particles, on the other hand, when given the identical mechanical disintegration treatment and spread out, give a non-self-supporting discontinuous film or coating which has little or no obvious utility.

PREPARATION

The essential step in converting the precursor chrysotile mineral silicate into the microcrystalline form is a chemical "etching" pretreatment whereby a selective removal of $Mg(OH)_2$ from the surfaces of the fibrils is effected and the creation of an outer residual sheath of hydrated silica with exposed reactive silanol groups* is formed (Fig. 4.3). This is verified by the corresponding increase in the ratio of the SiO_2/MgO over the starting chrysotile. Dissolution of the $Mg(OH)_2$ fraction at the surface of the fibrils leads to a reduction in the pH from a pH of 9 for the original chrysotile precursor to a pH of 7 for the microcrystalline mineral silicate. The magnesia content of microcrystalline silicate is approximately 37 percent as compared with 40 percent, the normal value associated with the chrysotile from which it is prepared. In other words, the chrysotile precursor has a higher degree of alkalinity than the microcrystalline silicate.

Treatment with the appropriate acidic medium increases the weight ratio of SiO_2/MgO by about 5 to about 30 percent, preferably 10 to 25 percent, as compared to that of the parent chrysotile without substantially altering the actual crystal structure of the chrysotile. Accordingly, a chrysotile having

$$\overset{\text{O}}{\overset{\|}{}}$$

*The formula for the silanol group is $-Si-OH$.

a weight ratio of SiO_2/MgO of 1:1 after treatment with the acidic medium will have a weight ratio of SiO_2/MgO of between about 1.05:1 and about 1.30:1 (69).

Electron microscopy reveals that the treatment weakens the fibrils by etching their surfaces so that they can be more readily reduced to submicron size by mechanical disintegration, something which is difficult to do with the untreated, extremely resilient fibrils. Furthermore, the treatment also serves to disperse at least some of the fibers into much thinner fibril aggregates. After the treatment severe mechanical attrition must be used so that at least about 5 wt percent of particles whose maximum dimension is under 1 μm are produced. The resultant product, when mixed with water, glycerol, glycols, or similar highly polar liquids, gives smooth, highly thixotropic gels at concentrations of the order of 1 to 3 percent. The particle-size distribution of the remaining 95 percent of the treated chrysotile will affect the rheology of the gels. For example, if the major portion of the remainder has a particle size not greater than 10 μm, the gel will be smoother than a gel formed using a product in which the major portion of the remainder has a particle size not greater than 30 μm. Ideally, a product in which the particles are substantially all under 1 μm in maximum dimension would be unusually smooth, but such a particle-size homogeneity is not easy to achieve.

The most generally accepted formula for chrysotile is $3MgO \cdot 2SiO_2 \cdot 2H_2O$. Chrysotiles from the various world supplies exhibit ratios varying from about 0.92:1 to about 1.055:1 depending upon the specific source and natural extraneous impurities. Literature indicates that the SiO_2 may constitute from 37 to 44 wt percent and the MgO may constitute from 39 to 44 wt percent of the chrysotile.

To produce the required change in chemical composition, weak or strong acids may be used, preferably at controlled elevated temperatures. For

fig. 4.3 *Microcrystalline-mineral-silicate microcrystal surfaces.*

example, 0.2 N hydrochloric acid may be used to treat chrysotile at about 5 to 10 percent solids for $\frac{1}{2}$ to 4 h at reflux. This treatment will produce an optimum increase in the SiO_2/MgO ratio of about 20 percent which maximizes rheology control in polar media. The use of a pressure digester to permit digestion under pressure permits a reduction in the time of treatment as well as in the concentration of acid required to effect a change to the optimum SiO_2/MgO ratio.

Satisfactory results have been obtained with hydrochloric acid; sulfuric acid; nitric acid; an acetylating mixture consisting of acetic acid, acetic anhydride, and trace amounts of sulfuric acid (as used in preparing cellulose acetate); and phosphoric acid. Salts which hydrolyze to produce hydrogen ions are also satisfactory and include such salts as the alkali-metal acid phosphates and sulfates. It is important that the acid not be too concentrated. For example, 0.4 N sulfuric acid at reflux will remove far too much MgO in 5 min, which causes an excessive change in the SiO_2/MgO ratio together with a rapid loss of yield and gelling properties. On the other hand, an acetylating mixture of 600 ml acetic acid plus 110 ml acetic anhydride plus 3 ml of concentrated sulfuric acid can be used safely even after an hour providing good control is kept to prevent the reaction from going beyond the desired yield and the desired SiO_2/MgO ratio.

A detailed procedure (69) for preparing microcrystalline silicate follows.

Five pounds by weight of chrysotile (SiO_2/MgO, 0.99:1) is mixed with 95 pounds by weight of 0.2 N hydrochloric acid. The product is divided into four equal portions which are heated at a boil under reflux. A portion of the product is removed at the end of 5 min, 15 min, 60 min, and 4 h to provide samples A, B, C, and D, respectively.

Utilizing like conditions, sample E is prepared by the use of 0.4 N HCl and heating for 5 min. Sample F is prepared by treatment with boiling 0.2 $N\,H_2SO_4$ for 5 min, and sample G is prepared by the use of 0.4 $N\,H_2SO_4$.

The yield and the SiO_2/MgO weight ratio of each product prepared in this way are reported in Table 4.1. When portions of the attrited samples are well dispersed at 3 percent solids in ethylene glycol, apparent viscosities

table 4.1 (69)

Sample	Yield, %	SiO_2/MgO	Viscosity, cP
A	87.5	1.19	27,300
B	87.9	1.16	34,200
C	88.4	1.22	36,300
D	87.2	1.24	38,000
E	78.0	1.54	16,000
F	88.2	1.24	31,400
G	76.6	1.48	2,400

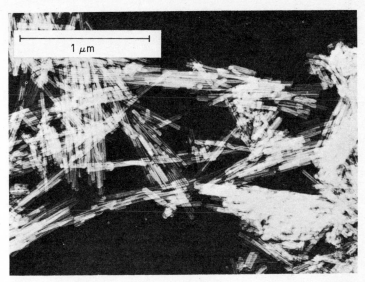

fig. 4.4   Chrysotile-mineral-silicate microcrystals.

obtained, measured in a Brookfield viscometer at 25 °C using a TB 10 spindle at 10 r/min, are also listed in Table 4.11.

Since chrysotile is slightly alkaline, the amount of acid needed will vary somewhat with the percent of chrysotile being treated.   Hence, if 10 percent or more solids are being treated with acid, somewhat higher concentrations are needed.

Proper high-shear mechanical attrition or mechanical disintegration is critical in producing an adequate proportion of submicron particles necessary to produce stable suspensions in polar media (Fig. 4.4).   Ball-milling the microcrystalline silicate leads rapidly to a product in which essentially all of the microcrystals become less than 1 μm in length (Fig. 4.5).

Other equipments that are preferred for handling high-solids pastes, and that also are more efficient in building up submicron particles with all the various microcrystal polymer products, are the Cowles Hi-Shear Mixer, such as Model 1-VG (Cowles Dissolver Company, Incorporated, Los Angeles, California), or the Rietz Extructor, also capable of effecting high shear during the mechanical disintegration (Rietz Manufacturing Company, Santa Rosa, California).

The acid pretreatment in combination with proper mechanical disintegration, in addition to making it possible to produce microcrystalline colloidal material of the original crystal structure, has still another advantage in that it removes impurities such as iron oxide and other acid-soluble impurities, which are present in a few percent in almost all chrysotile.   This purification,

fig. 4.5　Ball-milled microcrystalline silicate.

which takes place as part of the required acid pretreatment, has the advantage of improving the electrical properties of the microcrystalline mineral silicate as well as its whiteness.

The acid-pretreated chrysotile can be readily converted to almost 100 percent submicron particles by the appropriate dry or wet ball-milling of the product.

PROPERTIES

Chemical Composition and Physical Properties

Table 4.2 lists representative data for a developmental sample of a microcrystalline silicate.

An x-ray diffraction profile for microcrystalline silicate having an SiO_2/MgO ratio of 1.10:1.00 is given in Fig. 4.6.

The final viscosity of the microcrystalline colloidal silicate dispersions is highly dependent on the method of preparation. By slight adjustment of the preparation variables, such as type of dispersing equipment, time, and

*table 4.2   Composition and Physical Properties
of Microcrystalline Silicate*

Composition, wt %:
  Mg as MgO.. 37
  Si as SiO_2 43
  Water (700°C ignition)*................................... 12
Properties:
  Density (helium method), g/cm^3 .... 2.2
  pH of water slurry (1 wt %) 7
  Refractive index 1.5
  Surface area (N_2 adsorption), m^2/g† 70
  Adsorbed water, wt %:
    Typical product.. 1.5
    Equilibrium at 50% relative humidity and 25°C....... 2.0

*Water lost at 100°C is considered to be adsorbed, whereas water lost at temperatures of about 650 to 700°C is water of composition in the crystal lattice.

†Surface area quoted is for one grade. This can be varied, depending on the degree of mechanical treatment. The surface area of each of the samples was determined by the conventional nitrogen adsorption method using the Perkin-Elmer adsorptometer.

mixing velocity, it is possible to modify at will some of the properties of a given formulation. Dispersions comprising microcrystalline colloidal silicates possess extreme non-Newtonian properties. As a result, the apparent viscosity of such systems drops very rapidly under conditions involving high-speed mixing. On the other hand, of course, the higher the buildup of submicron microcrystalline colloidal silicate particles in the composition, in general, the higher the viscosity of the composition at a stationary state.

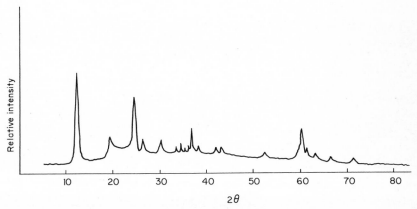

fig. 4.6   X-ray diffraction pattern of microcrystalline silicate (Avibest, SF).

table 4.3 (69)

Liquid	MCS, %*	Silica, %*	Viscosity, cP, at spindle speeds, r/min, of		
			0.6	3.0	12.0
Mineral oil	1.0	0.0	66,400	18,300	5,080
	2.0	0.0	312,000	119,500	33,330
	0.0	2.0	7,000	2,150	875
	0.0	4.0	349,000	120,000	36,700
Ethylene glycol	2.0	0.0	99,900	33,300	10,300
	4.0	0.0	465,000	193,000	68,300
	0.0	8.0	100	120	100
	0.0	15.0	300	300	270
Water	2.0	0.0	4,200	1,300	400
	4.0	0.0	81,000	18,000	5,300
	0.0	4.0	100	50	30
	0.0	8.0	700	200	350

*All dispersions were first prepared by stirring with a glass stirring rod driven by a Cenco stirring motor at 1,200 r/min for 30 min.

In Table 4.3 the efficiency of the microcrystalline colloidal silicate (MCS) as a thickening agent in various liquids is compared to commercially available submicroscopic pyrogenic silica (69).   Apparent viscosities were obtained using a Brookfield viscometer.

Similar results are demonstrated with pigmented coating compositions (paints) in Table 4.4.

The above data demonstrated that the microcrystalline colloidal silicate is more effective on a weight basis than the commercially available silica thickener.

table 4.4 (69)

Coating system*	MCS, %†	Silica, %†	Viscosity, cP, at spindle speeds, r/min, of		
			1.0	10	100
Chrome oxide green, linseed oil	0.0	0.0	56,000	7,600	1,380
	0.0	1.0	58,000	8,000	1,440
	1.0	0.0	120,000	18,400	3,400
Red iron oxide, alkyd resin	0.0	0.0	2,200	420	192
	0.0	1.0	2,800	540	220
	1.0	0.0	4,000	700	264

*Systems prepared by milling in a ball mill until a grind of 6 to 7 was obtained.
†Based on the binder content of coating.

Rheology of Microcrystalline Silicate in Organic Liquids

Relatively low levels of microcrystalline silicate dispersed in organic liquids cause pronounced increases in the viscosity of the fluid at low shear rates. As the shear rate increases, this tendency is diminished. Ferraro (70) has investigated the rheological behavior of such dispersions by using rotational viscometry measurements. He found that microcrystalline silicate gels exhibit properties that are remarkably similar to those reported for microcrystalline cellulose gels (3,4,23,24).

The measured shear-stress rate of shear data could be related by the power law (70):

$$\tau = \tau_y + k\dot\gamma^n$$

where τ = shearing stress, dynes/cm^2
τ_y = estimated yield value, dynes/cm^2
k = apparent plastic viscosity $(\tau - \tau_y)$ at unit shear rate
$\dot\gamma$ = rate of shear, sec^{-1}.
n = pseudoplastic index (<1)

The index of thixotropy, B, as defined by Weltmann (71), was also calculated for the systems at two different shear rates. By definition, B is the rate of change of shearing stress with the logarithm of time at any given shear rate.

Organic liquids that were investigated by Ferraro included tricresyl phosphate, dioctyl phthalate (di-2-ethylhexyl phthalate), and a silicone fluid. The results of the study are summarized in the following tables.

table 4.5   Rheology of Microcrystalline Silicate-Tricresyl Phosphate*
Systems at 25.0°C—Haake SV I Couette System (70)

	Level of microcrystalline silicate, wt %			
Value	0.25	0.50	1.00	1.50
n	0.69	0.56	0.51	0.53
rpm = unit shear rate	1.14	1.16	1.17	1.16
τ_y, dyn/cm^2	31.8	153.0	331.0	509.0
k	5.7	21.9	39.5	56.0
η, P:				
1 s^{-1}	37.5	175.0	371.0	565.0
10 s^{-1}	6.0	23.2	45.9	69.9
100 s^{-1}	1.7	4.4	7.3	11.6
B, dyn/cm^2:				
At 10.8 r/min	11.0	52.0	74.0	227.0
At 97.2 r/min	20.0	34.0	104.0	233.0

*The tricresyl phosphate (FMC Kronitex AA) is Newtonian and has a specified viscosity of 1.20 P at 20°C.

table 4.6 Rheology of Microcrystalline Silicate-Di-2-ethylhexyl Phthalate* Systems at 25°C—Haake SV I Couette System (70)

Value	Level of microcrystalline silicate, wt %			
	0.25	0.50	1.00	1.50
n	0.68	0.49	0.38	0.32
rpm = unit shear rate	1.14	1.17	1.18	1.19
τ_y, dyn/cm^2	6.4	20.4	51.0	96.0
k	6.7	36.9	125.0	242.0
η, P:				
1 s^{-1}	13.1	57.3	176.0	338.0
10 s^{-1}	3.9	13.4	35.1	60.1
100 s^{-1}	1.6	3.8	7.6	11.3
B, dyn/cm^2:				
At 10.8 r/min	7.0	41.0	108.0	185.0
At 97.2 r/min	8.0	31.0	104.0	191.0

*The DOP (FMC Corporation) is a Newtonian fluid and has a specified viscosity of 0.80 P at 20°C.

The foregoing data illustrate how microcrystalline silicate exhibits an unusual capability to impart pseudoplastic character to such liquids. The presence of a yield value as well as thixotropy strongly suggests that the dispersed microcrystalline silicate is not only hydrogen bonded to the organic-liquid molecules but also organized in the dispersion as a "house-of-cards" or brush-heap structure. The presence of the yield value is responsible

table 4.7 Rheology of Microcrystalline Silicate-Plastisol* Systems at 25°C—Haake SV II Couette System (70)

Value	Level of microcrystalline silicate, wt %		
	0	0.56	1.11
n	0.97	0.81	0.94
rpm = unit shear rate	1.10	1.11	1.10
τ_y, dyn/cm^2	0	0	255.0
k	56.0	170.0	113.0
η, P:			
1 s^{-1}	56.0	170.0	368.0
10 s^{-1}	51.8	110.0	123.0
100 sec^{-1}	47.9	70.9	88.4
B, dyn/cm^2:			
At 10.8 r/min	32.0	131.0	150.0
At 97.2 r/min	213.0	372.0	477.0

*GEON 120 × 203 (vinyl chloride resin) combined with DOP (dioctyl phthalate).

table 4.8 Dispersion of 0.5 wt % Microcrystalline
Silicate into Polyester Resin with a Cowles Dissolver (72)*

Mixing time, min	Brookfield viscosity at 25°C, P		
	1.0 r/min	2.5 r/min	20 r/min
0	1	1	1
5	126	56	12
15	256	119	26
25	297	129	28
40	303	136	31

*P. 400.

for the fact that the dispersions do not show any settling tendencies with time when undisturbed.

Rheology-Control Agent
for Polymer Systems

Strunk (72) has reported on the application of microcrystalline silicate as a highly effective rheological-control agent for unsaturated polyester resins. The viscosity behavior of unsaturated polyester resins is essentially Newtonian; that is, the viscosity does not change with the rate of shear. In hand lay-up or spray–hand lay-up contact-molding applications it is necessary to add a rheological-control agent to prevent undesirable drainage of the polyester from vertical surfaces. Pseudoplastic polyester dispersions utilizing only 0.5 wt percent microcrystalline silicate were prepared by using an efficient dispersing tool, such as a Cowles dissolver. The data are reproduced in Table 4.8.

It was also observed that an increase in viscosity is obtained by the addition of polyols to a styrene-based polyester dispersion containing microcrystalline silicate. The addition of 0.25 wt percent glycerol to a polyester containing 0.50 wt percent microcrystalline silicate causes a 50 percent increase in the viscosity values. These effects are described in Table 4.9. As a part of this study it was found that the maximum amount of a polar additive must be determined in each case. The data in Table 4.10 clearly show this effect for glycerol on dispersions of microcrystalline silicate in three polyesters (resins A, B, and C).

Degree of Bonding of
Microcrystalline Silicate
to Polyhydroxylic Liquids

Strong bonding forces exist between microcrystalline silicate and such polyhydroxylic compounds as glycerol or glycols, which can cross-link the microcrystals much more effectively. The pseudoplasticity of dispersions of un-

table 4.9 Effects of Several Polyols on the Viscosities of Dispersion of 0.5 wt % Microcrystalline Silicate in a Polyester Resin* (72)†

Polyol		Brookfield viscosity at 25°C, P		
Type	Amount, wt %	1.0 r/min	2.5 r/min	20 r/min
None	0.0	170	83	18
Ethylene glycol	0.25	180	85	18
Diethylene glycol	0.25	210	98	25
Glycerol	0.25	247	116	26
Pentaerythritol	0.25	149	72	15

*American Cyanamid Laminac 4123.
†P. 401.

table 4.10 Effect of the Concentration of Glycerol on the Viscosities of Dispersions of Microcrystalline Silicate in Several Commercial Resins* (72)†

Avibest-C, wt %	Glycerol, wt %	Brookfield viscosity at 25°C, P		
		1.0 r/min	2.5 r/min	20 r/min
Resin A, no cobalt				
0.00	0.00	1	1	1
0.25	0.00	62	30	8
0.25	0.25	114	50	12
0.25	0.50	167	98	19
0.25	0.63	49	24	8
0.25	1.00	16	10	3
Resin B, cobalt promoter				
0.00	0.00	2	2	2
0.25	0.00	29	16	5
0.25	0.10	42	20	5
0.25	0.30	190	87	23
1.00†	0.30	99	49	14
0.25	0.50	174	85	25
Resin C, cobalt promoter				
0.00	0.00	1	1	1
0.25	0.25	44	20	5
0.25	0.50	122	61	16
1.00‡	0.50	105	48	12

*Styrene-based general-purpose unsaturated polyester.
†P. 401.
‡Pyrogenic silica.

saturated polyester prepared from microcrystalline silicate that has first been blended with diethylene glycol in the weight ratio of 1/1 or 1/2 is far greater than that of the dispersion prepared by adding the glycol directly to the dispersing medium or even to the final dispersion containing the microcrystalline silicate.

This synergistic effect is evident over long time periods. The viscosity of certain dispersions has been periodically measured over a period of several months with no observable decrease in viscosity at various shear rates (70).

In the blend of microcrystalline silicate with glycol, the glycol is tenaciously bonded because of the available silanol groups on the surfaces of the microcrystals. Dispersion of this blended product in various liquids apparently occurs with little effect on the degree of bonding between the glycol and the microcrystalline silicate. Table 4.11 contains data that show the synergistic effects described above, and Table 4.12 contains data on the long-term stability of the rheological characteristics of dispersions containing the blend. In these illustrations the blends were prepared with a commercially available liquid-solid twin-cone blender.

table 4.11 Effects on Viscosity of the Mode of Addition of Diethylene Glycol to a Ternary System* (70)†

	Brookfield viscosity at 25°C, P		
Mode of glycol addition	1 r/min	2.5 r/min	20 r/min
Blended onto microcrystals	70	31	8
Dissolved into polyester	43	20	5

*Consisting of microcrystalline silicate (0.25 %), glycol (0.50 %), and general-purpose polyester (1 P, 99.25 %). The system was promoted before measurement of the viscosity.
†P. 4.

table 4.12 Brookfield Viscosity with Time* (70)

	Brookfield viscosity at 25°C, P		
Time, months	1 r/min	2.5 r/min	20 r/min
0	34	19	4
1	41	21	7
2	42	21	7
3	46	24	7

*The system consisted of microcrystalline silicate (0.25 %), diethylene glycol (0.50 %), and general-purpose polyester (1 P, 99.25 %). The glycol was blended with the microcrystals, and this blend was then dispersed into the polyester. The system was promoted prior to storage for the long-term study.

Role of Dimethyl Sulfoxide (DMSO) in Opening Up Chrysotile and / or Microcrystalline Silicates (73)

When chrysotile asbestos fibers or microcrystalline silicates made from them are soaked in dimethyl sulfoxide or mixtures of dimethyl sulfoxide and water and subsequently beaten in a simple pulping device such as a Bauer refiner, the aggregated fibrils are more readily dispersed to individual fibrous particles (73).

Asbestos fibers so deaggregated may have the DMSO readily washed from them and now are in an especially favorable state—both in terms of increased surface area and fibril-surface hydration—to be converted into the microcrystalline colloidal form by means of the controlled HCl treatment previously described (69).

Morphological Studies of Chrysotile and Microcrystalline Silicates Produced Therefrom

In the course of our studies of the conversion of chrysotiles into colloidal microcrystalline silicates, we undertook exhaustive electron-microscopy studies.

Normal chrysotile fibrils are long, uniformly slender fibrils as shown in Fig. 4.1.

The process of dissolving $Mg(OH)_2$ from the surfaces of the fibrils, which leads to an increase in the SiO_2/MgO ratio, leaves the surfaces of the fibrils with a uniform surface of hydrated silica containing silanol groups (Fig. 4.1).

Severe mechanical attrition of the acid-pretreated chrysotile, preferably at high solids (20 to 30 percent) in a Readco-type extruder, leads to a buildup of submicron rodlike particles. Ball-milling the precursor microcrystalline silicate produces an even higher percentage of submicron and still shorter-length particles as indicated by Fig. 4.5.

The hydrated silica surface of microcrystalline silicate microcrystals was found to have an unusual affinity for spherical colloidal silica particles having diameters of about 100 to 150 Å. This phenomenon, in which the colloidal spherical particles appear to be adsorbed uniformly on the surface of the fibrils, is clearly shown by Figs. 4.7 and 4.8.

Figure 4.9 illustrates an interesting morphology of chrysotile fibrils; careful examination of the surfaces of several fibrils reveals a central core or channel running the length of each fibril.

In taking electron micrographs of chrysotile fibrils or microcrystalline silicate particles produced therefrom, caution must be exercised to prevent the electrons from heating the fibrils in excess of about 600°C. When this occurs, the approximately 10 to 12 percent water of hydration bound within

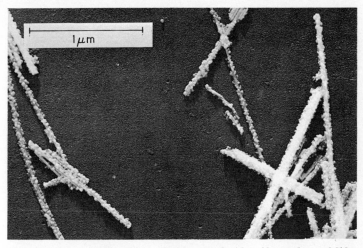

fig. 4.7   *Microcrystalline silicate particles coated with uniform spheres of SiO$_2$.*

the SiO$_2$–MgO lattice is dislodged as steam and an attendant disruption of the original morphology takes place as is clearly evident in Fig. 4.10.   If this precaution is not taken, serious misinterpretations of related electron micrographs may well occur.

APPLICATIONS

Although microcrystalline silicates are not commercially available, they have been explored as developmental products.   For example, they have been used

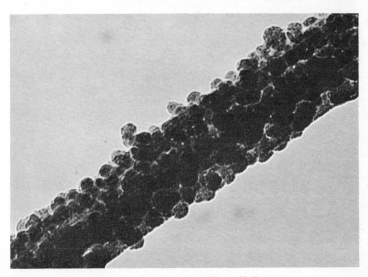

fig. 4.8   *Silica spheres on a single mineral silicate fibril.*

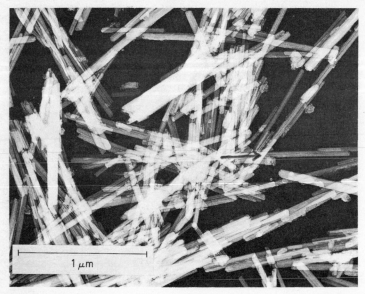

fig. 4.9 *Channels or cores in chrysotile fibers.*

fig. 4.10 *Mineral silicate fibrils showing decomposition by overheating in electron microscope.*

in unsaturated polyester and epoxy resins to control their flow properties. Figure 4.11 compares the contribution of microcrystalline silicate to the rheology of a polyester resin with much higher concentrations of pyrogenic silica. The polyester resin used is a maleic anhydride–phthalic anhydride–propylene glycol condensation product dissolved in styrene.

Microcrystalline colloidal silicates have also been found especially useful for the clarification and densification of sewage sludge when used in dry flour or powder form. Because of its absorptive properties for dyes and phenolic compounds, other industrial waste waters such as dyehouse waste may be effectively treated with this form of modified chrysotile.

They are excellent ingredients for column separation and have been effective when used this way for decolorizing a solution of brown sugar. Microcrystalline colloidal silicates in dry powder form can be used to filter gases

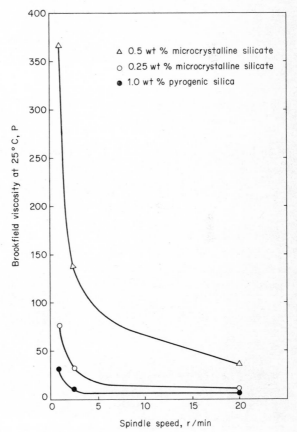

fig. 4.11 Brookfield viscosity of dispersions of microcrystalline silicate in polyester resin (72).

such as cigarette smoke because of their unique surface affinity for phenol-based chemicals and other organic compounds. For example, they can be an effective filter material for removal of benzopyrene in cigarette smoke. Similarly, when added to solutions or dispersions of dyes, they are capable of decolorizing dye solutions. Gels of this product were found to be quite effective in producing stable graphite greases, preventing any separation of the graphite particles on standing for long periods of time so that their bodying action can be utilized effectively to stabilize suspensions of particulate matter such as pigments, etc. The compatibility and stability of these gels in strong acids as well as in acid-salt solutions such as $ZnCl_2$ make them candidates as long-lasting gelling media for leakproof dry-cell batteries.

Some chrysotiles as mined contain significant amounts of spherical particles of magnetite. Conversion of such raw materials to the colloidal microcrystalline state can be done in such a way as to leave particles of magnetite on the submicron fibril surfaces or on the surfaces of aggregates thereof. When these rodlike particles containing magnetite are uniformly dispersed in a fluid such as an unsaturated polyester resin and the mixture is exposed to a permanent magnetic or an electromagnetic field, useful thickening effects such as those shown in Fig. 4.12 are observed. By the use of an alternating electromagnetic field, an "on-off" rapid sequence of extreme thickening and thinning action can be controlled.

The melting point of aluminum (660°C) coincides closely with the temperature at which chrysotile fibrils and/or microcrystalline silicate fibrils lose their bound water of hydration. This coincidence has been used to develop a new way of producing foamed aluminum and other low-melting foamed metals or mixtures thereof (74,75).

fig. 4.12 *Polyester resins containing a dispersion of microcrystalline mineral silicate plus magnetite particles.*

Inasmuch as some metals such as lead, fluxes, etc., melt well below 660°C, the use of uniformly dispersed colloidal microcrystalline silicate particles within the matrix of such low-melting metals and fluxes also has been advanced (75). The concept here is that the deaggregated, individually dispersed colloidal rodlike particles of microcrystalline silicate can provide whisker-like reinforcement of the properties of these low-melting metal compositions.

A chart showing the projected potential uses of colloidal microcrystalline mineral silicates is given in Table 4.13.

table 4.13   Potential Microcrystalline Silicates Uses

Use	Examples
Filters	Liquids, gases
Suspending agent	Paint strippers, metal cleaners
Binder	Catalysts, abrasive wheels
Viscosity control	Plywood adhesives
Reinforcement	Plastics, metals
Dispersion hardening	Metals
Nucleation agent	Cloud seeding
Catalyst	Combined with other microcrystalline products

CHAPTER FIVE

Microcrystalline Amyloses

INTRODUCTION

Amylose is a linear polymer of 1,4-anhydroglucose units linked through α-glycosidic bonds, whereas cellulose is an identical polymer linked through β-glycosidic bonds. Chemically, both of these carbohydrate polymers analyze to be $C_6H_{10}O_5$. Natural starches consist of a mixture of amylose and branched amylopectin molecules, with a higher proportion of the starch consisting of branched amylopectin. In more recent years hybrid starches have been developed in which the proportion of amylose to amylopectin has been increased substantially. Figure 5.1 compares the structures of amylose and amylopectin, respectively.

Amylose in relatively pure form must be recovered by purification procedures (usually fractionation) from the natural or hybrid forms of starch, respectively. Per se, amylose or starches rich in amylose are not capable of forming stable gels or dispersions in water even after considerable mechanical disintegration.

In our investigations of microcrystalline amyloses (76), our starting starch raw material was a special developmental grade containing approximately 80 to 85 percent of crystalline amylose.

Naturally synthesized amylose is not as highly crystalline as cellulose and somewhat more readily attacked by acid. It is especially receptive to acid

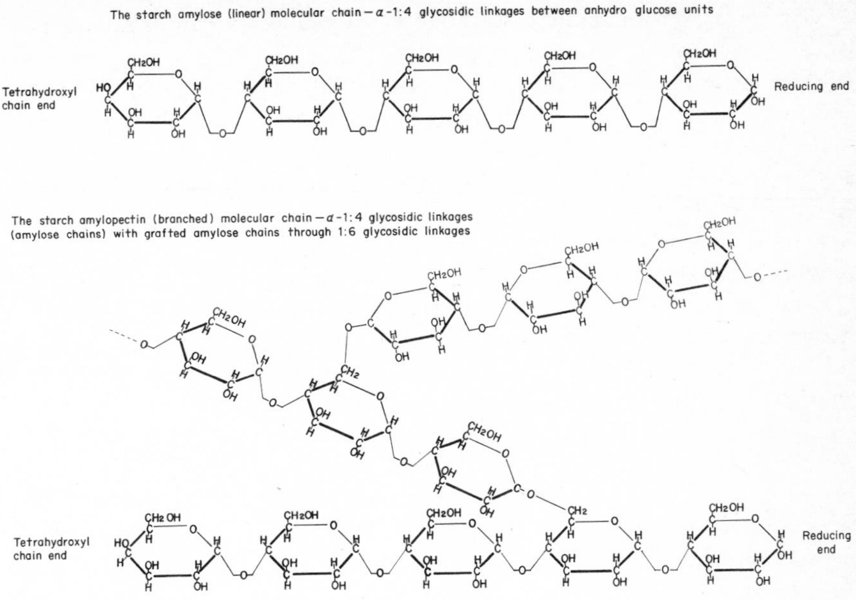

The starch amylose (linear) molecular chain $-\alpha-1:4$ glycosidic linkages between anhydro glucose units

The starch amylopectin (branched) molecular chain $-\alpha-1:4$ glycosidic linkages (amylose chains) with grafted amylose chains through 1:6 glycosidic linkages

fig. 5.1 *The starch amylose: linear molecular chain of 1,4-α-glycosidic linkages between anhydroglucose units.*

attack paralleling the technology developed for microcrystalline celluloses; milder conditions of hydrolysis are required, and purification treatments are somewhat more tedious. Nevertheless, microcrystalline amylose when isolated and converted into stable suspensoids exhibits many of the unique rheological properties and viscosity-stability properties in the presence of heat of microcrystalline cellulose suspensoids. One of the fundamental differences, of course, is that microcrystalline amylose has 4 cal/g, whereas microcrystalline cellulose has essentially 0 cal/g.

PREPARATION

The starting raw material from which microcrystalline amyloses may be prepared should be preferably a starch containing a major proportion of amylose. Reasonably high yields are obtained with a raw material containing about 85 percent amylose. Yields necessarily decrease in direct proportion to the decreasing amylose content of the starting raw material.

Treatment of high-amylose starch to convert it to its level-off degree of polymerization (LODP), as with cellulose, may be carried out using various inorganic or organic-acid media. The object is to dissolve away regions of low lateral order so that the water-insoluble, highly crystalline residue may be converted into a stable suspensoid by subsequent vigorous mechanical shearing action.

As a preferred procedure the hydrolysis of the high-amylose starch raw material is best carried out by treating it with an aqueous solution of hydrochloric acid at a concentration from at least 0.01 N up to 2.5 N for from 1 to 30 min under reflux conditions. Acid concentrations over 0.25 N result in much poorer yields, while concentrations under 0.1 N require excessive time for sufficient hydrolysis. The acid-hydrolyzed product is generally thoroughly washed to remove acid prior to attrition.

The mechanical disintegration of the amylose material may be carried out in several ways: by subjecting it to attrition in a mill, to a high-speed cutting action, or to the action of high pressures. Disintegration is generally carried out in the presence of a liquid medium, preferably water, although, where high-pressure attrition alone is employed, such a medium while desirable is not necessary. Although water is the preferred liquid medium, other liquids are suitable. Sugar solutions, polyols, of which glycerol is an example, and alcohols, particularly ethanol, isopropanol, and the like, are good examples of suitable liquid media. Whatever method is used, the disintegration is carried out to such an extent that the resulting finely divided product is characterized by its ability to form a stable suspension in the liquid medium in which it is attrited or in which it is subsequently dispersed. By a stable suspension or dispersion is meant one from which the attrited material will contain sufficient submicron particles to prevent settling and promote indefinite suspension, even for periods measured in terms of weeks or months. In general, dispersions of the attrited amylose persist up to at least about 5 percent solids, while the gel phase begins at about 10 percent solids in an aqueous medium.

The following example defines typical conditions for producing microcrystalline amylose from an amylose raw material containing approximately 85 percent amylose (76). As will be noted, hydrolysis must be continued for a sufficient length of time so that a leveling off of the hydrolysis reaction is attained before stable smooth pseudoplastic suspensoids can be made from the hydrolysis-resistant residue.

In the following experiment 600 ml of 0.1 N hydrochloric acid were heated to boiling in each of four closed reaction flasks. Fifty grams of dry starch (85 percent amylose) were added to the boiling hydrochloric acid in each of the reaction flasks. Acid hydrolysis in each flask was continued for 1-, 5-, 10-, and 15-min periods, respectively.

After each hydrolysis the product was quenched in an ice bath, washed free of acid, and vacuum-dried at 50°C. After a yield of the material was calculated, a slurry of 5 percent solids in water was attrited in a Waring Blendor for 15 min. The dispersions in which hydrolysis was continued for 1, 5, and 10 min, respectively, settled out of solution in varying degrees after attrition in the Waring Blendor for 15 min. The dispersion of the product which was hydrolyzed for 15 minutes was a white, stable dispersion after attrition.

The yield of degraded amylose which received the 15-min hydrolysis and which subsequently produced the stable dispersion was 76 percent.

PROPERTIES AND APPLICATIONS

Except for the previously mentioned differences in caloric values, microcrystalline amyloses and microcrystalline celluloses have many similar properties and applications. Inasmuch as microcrystalline cellulose has been produced for commercial uses on a worldwide basis for some years whereas microcrystalline amyloses are still in the development stage, reference to the chapter on microcrystalline celluloses might prove useful in projecting possible new applications for microcrystalline starches.

Figure 5.2 illustrates a moldable 12 percent aqueous gel of microcrystalline amylose prepared from an 85 percent amylose starch raw material. Its appearance, spreadability, fat-like characteristics, rheological properties, etc., parallel those of microcrystalline cellulose.

Microcrystals of amylose at their LODP exhibit a beaded structure (Fig. 5.3a) when recovered from a 10 percent suspensoid that had been attrited in a Waring Blendor for 30 min. Figure 5.3b shows ultrasmall unit particles (100 to 200 Å) recovered from the same suspensoid after more severe treatment—60 min in a blender at high speed.

A most interesting lamella morphology was observed when amylose molecules in solution were allowed to crystallize (Fig. 5.4). The amylose molecules appear to crystallize out of solution by chain folding; the thickness of the lamellae (circa 100 Å) corresponds to the repeat length in the chain fold. Iodine is not absorbed at all by this morphological form of crystalline starch.

Although the x-ray diffraction patterns for amylose raw material (Fig. 5.5) and microcrystalline amylose derived therefrom (Fig. 5.6) clearly show an increase in lateral order for the latter form, the levels of lateral order are

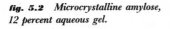
fig. 5.2   Microcrystalline amylose, 12 percent aqueous gel.

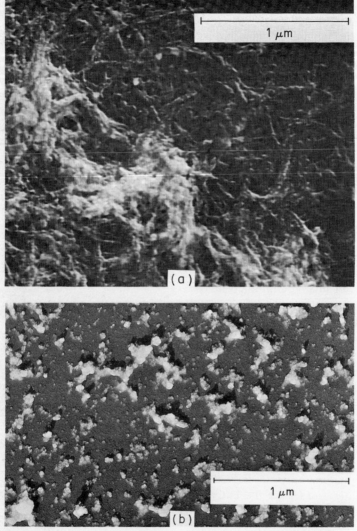

fig. 5.3 (a) Amylose microcrystals; (b) amylose microcrystals recovered from the supernatant diluted suspension (0.1 percent) after attriting a 10 percent suspensoid in an Osterizer for 60 min at 25°C.

not as high as for their cellulose counterparts. Figure 5.7 gives the x-ray diffraction pattern for the recrystallized lamella form of amylose.

Figure 5.8 describes the rates of structural recovery of the yield points with time for a microcrystalline amylose gel—behavior not unlike that observed for suspensoids of microcrystals produced from other natural and/or synthetic linear polymer precursors (77).

fig. 5.4   *Recrystallized amylose lamellae microcrystals.*

Aviamylose gels and suspensoids are smoother, shinier, and more translucent than pure microcrystalline cellulose gels. They hold special promise for future developments in the food and cosmetics industries. The fact that structural products made of microcrystalline amylose would be expected to be bioassimilable holds forth their possible future use in prostheses as well as in pharmaceutical applications.

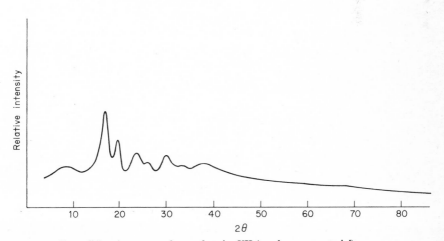

fig. 5.5   *X-ray diffraction pattern for amylomaize VII (amylose raw material).*

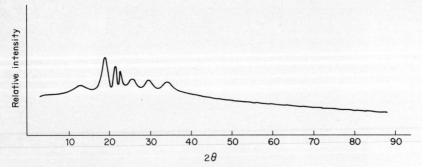

fig. 5.6 *X-ray diffraction pattern of microcrystalline amylose.*

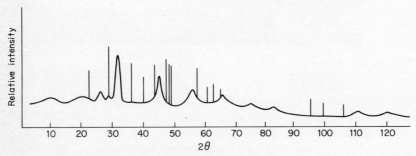

fig. 5.7 *X-ray diffraction pattern of recrystallized amylose.*

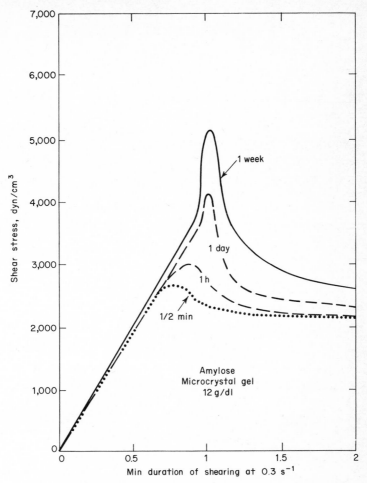

fig. 5.8 *Progression of yield-stress buildup after agitation of aqueous microcrystalline amylose gels (77).*

PART TWO

Man-Made Polymers

CHAPTER SIX

Microcrystalline Polyamides

INTRODUCTION

Microcrystalline nylons are prepared from their corresponding polyamide precursors by selective chemical treatments. These treatments are designed to cleave and remove regions of lowest lateral order in the more accessible morphological "cracks" or hinges within each precursor polyamide matrix.

As is uniformly required for each member of the microcrystal polymer family, conditions of pretreatment (usually acidic for polyamides) must be selected that preclude excessive swelling of the constituent crystallites. Appropriate subsequent mechanical disintegration of the unhinged microcrystals produces a family of dispersions of varying particle-size distributions and rheological properties.

Two types of polyamides have received the most extensive study for conversion into microcrystalline forms—those prepared by the polymerization of monoaminomonocarboxylic acid and those obtainable from diamines and dibasic carboxylic acids. Other polyamides also have been studied, such as, for example, the caprolactams, which have the crystalline-amorphous structure and are also satisfactory.

PREPARATION

The polyamide raw material must be a linear polyamide having a molecular weight or degree of polymerization sufficiently high so that it is capable of forming a self-supporting filament or fiber. Microcrystalline polymer products prepared from such polyamide precursors are characterized in their ability to form stable colloidal dispersions in liquid media.

Synthetic linear polyamides having molecular weights sufficiently high to allow conversion into fibers possess toughness and elastic characteristics which make grinding of them into finely divided particles very difficult. Precipitation of such polyamides from solution tends to produce stringy, cohesive precipitates. Lower-molecular-weight polyamides may have the necessary physical characteristics to allow the resin to be ground to a finely divided state and to be precipitated in a finely divided state. Regardless of what the molecular weight of these forms of finely divided polyamides is and of whether they are produced by a grinding operation or by a dissolving and precipitation procedure, the particles are characterized by the typical polyamide crystalline-amorphous network. They still do not produce the stable colloidal dispersions having the unique rheological properties of suspensoids of microcrystalline nylons produced by selective chemical pretreatments. Furthermore, mechanical grinding of a polyamide cannot produce single microcrystals in quantity having dimensions of the order of 100 to 150 Å and a high degree of crystallinity.

The fiber-forming synthetic linear polyamides, preferably in fibrous form, are converted to microcrystalline form by treatment with a dilute aqueous mineral-acid solution at a temperature of at least 50°C but below the melting point of the polyamide without substantial dissolution thereof. The acid treatment must not be so drastic as to excessively swell the crystalline portion of the polyamide or to destroy the microcrystals in the original morphological network but must be sufficient so as to selectively attack the interconnecting molecular chains constituting the amorphous hinges or cracks in the gross crystalline morphology. In the selective attack on the amorphous portions some of the amorphous material is dissolved and some of the molecular chains in the hinges are cut or weakened sufficiently so that upon mechanical disintegration, microcrystals are freed from the aggregated microcrystals.

Acids that are satisfactory to effect the selective treatment of the polyamides are, for example, hydrochloric acid, sulfuric acid, nitric acid, and phosphoric acid.

The preferred mineral acid is hydrochloric acid diluted with water to a concentration of less than 15 percent. Hydrochloric acid concentrations as low as 2 or 3 percent are useful but require much longer treatment periods and/or greater mechanical attrition of the partially degraded polymer. Preferably, dilute hydrochloric acid concentrations of about 5 to about 10 percent are used to obtain the most desirable results. Higher concentrations than this will greatly reduce the yield of gel-forming product because of progressive attack on the relatively acid-sensitive insoluble polyamide residue.

The length of time during which the linear polyamide is subjected to the dilute acid treatment will vary according to the concentration of the acid and, to some degree, the physical form of the polyamide employed. However, in general, from $\frac{1}{2}$ to about 6 h will produce satisfactory yields, the shorter periods being used with higher acid concentrations. For example, treatment of filament-forming linear polyamide with a 5 percent dilute aqueous hydrochloric acid solution at boiling for $2\frac{1}{2}$ h will yield a satisfactory product with an 89 percent yield, while a 10 percent dilute aqueous hydrochloric acid solution at boiling for 1 h provides a product yield of 80 percent.

Other dilute aqueous mineral acid solutions are useful in concentrations which will not dissolve or swell the microcrystals in the polymer structure but will selectively attack the amorphous areas thereof and produce a gel-forming product by mechanical shear. Dilute aqueous mineral acids which will degrade filament-forming linear polyamide to a degree of degradation about equivalent to that obtained by treatment thereof with about 5 percent aqueous hydrochloric acid solution at a temperature of at least 50°C for $\frac{1}{2}$ h are also applicable.

Aqueous and/or organic-solvent solutions of organic acids can be used, if desired, to effect the unhinging of the molecular chains in the amorphous regions and thereby make the aggregated microcrystals available for mechanical separation of individual microcrystals. For example, aqueous formic acid solutions of at least 50 percent concentration with suitable temperature and time conditions are satisfactory. Similarly, an ethanol solution of peracetic acid containing about 5 percent of the acid is also satisfactory.

For the purpose of preparing stable dispersions or gels of microcrystalline polyamides, the microcrystals may be treated with a swelling agent prior to attrition, or they may first be attrited and then dispersed with mechanical agitation in a liquid swelling medium.

The following procedure is an example of one method of preparing microcrystalline polyamides (78).

A 150-g quantity of staple polyhexamethylene adipamide fiber was hydrolyzed for 8 h at 72°C in 1,150 ml of 5 percent aqueous hydrochloric acid solution. After washing with water until most of the acid was removed, the hydrolyzed polyamide was subjected to attrition for 20 min in a Hobart mixer and then for 7 min in a Mixmaster in water at a solids concentration of 35 percent to reduce the product to a fibrous pasty mass. A 400-ml volume of a 90 percent aqueous formic acid solution was added to this fibrous mass, which was mixed and allowed to stand overnight at room temperature. The acid mixture was neutralized to pH 7 with sodium hydroxide, and the entire suspension was dialyzed overnight. After filtration, the product retained a great deal of water, and additional water was added to obtain a solids concentration of 10 percent. The mixture was attrited for 20 min in a Mixmaster to obtain a smooth, stable gel.

A portion of the above stable gel was recovered by solvent extraction using acetone. The untreated or virgin polyamide fiber precursor and a sample of the dried, attrited microcrystalline product were heated individually in a Perkin-Elmer recording differential scanning calorimeter calibrated to 0.1°. The virgin fiber exhibited a melting range of 245 to 268°C with a peak at 259°C in the melting curve. The microcrystalline residue showed two separate melting ranges, one in a range of 243 to 250°C and the second in a range of 252 to 264°C with a peak at 260°C.

Figure 6.1 is a flowsheet showing steps in producing microcrystalline polyamides from resins in pellet form.

PROPERTIES

Mechanical attrition of pastes of microcrystalline nylons is most effective at high solids (40 to 45 percent). When high shear energy is applied to such pastes and they are subsequently diluted, very high proportions of the discrete nylon microcrystals of the order of 100 to 200 Å in maximum dimensions disperse in water to form stable suspensoids (Fig. 6.2).

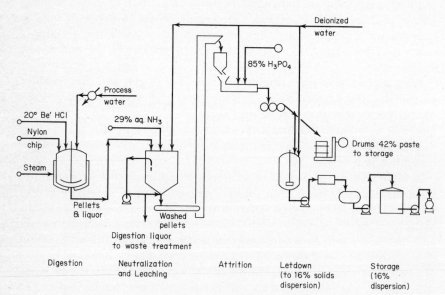

fig. 6.1   *Flowsheet for the preparation of microcrystalline polyamides.*

fig. 6.2   *Microcrystalline nylon, 15 percent aqueous suspensoid.*

The chemical composition and properties of such a high-solids paste made of polycaprolactam (nylon-6) are listed in Table 6.1.

Procedure for Dilution of High-Solids Pastes to Stable Suspensoids

The following procedure describes the preparation of stable suspensoids from concentrated pastes (40 to 45 percent) of microcrystalline nylons.

1. Prepare a 1,000-g slurry containing 5 wt percent solids from a 44 percent microcrystalline nylon paste with distilled or deionized water. Adjust the pH to 5 to 6 with dilute H_3PO_4, dilute HCl, or dilute NH_4OH depending on the pH of the paste.

2. Mix the slurry at room temperature for 5 min in a Waring Blendor at medium speed. Transfer the contents into a 1-liter graduated cylinder.

3. Allow the slurry to stand for at least 30 s at $25°C \pm 1°C$ without any agitation. Carefully siphon off the suspendable fraction present in the cylinder.

4. The solids content present in the suspendable fraction will be about 80 percent of the total solids present in the initial slurry.

The proportion of free microcrystals present in a dilute suspension of microcrystalline nylon is highly dependent on the mechanical equipment used and the concentration of solids at which the attrition of the paste takes place. Figure 6.3 demonstrates this point clearly. A Readco Extructor was used to produce samples having three broad levels of dispersibility—regular, medium, and severe.

The same high-solids paste was passed through the mill repeatedly to give an increasingly fine suspensoid or dilution to 10 percent solids.

Cranstoun and Cruz (79) developed a fractionation procedure to obtain fractions of aqueous dispersions of the microcrystalline nylon suspensoids containing homogeneous particle sizes.

table 6.1 *Properties of Microcrystalline Nylon-6 (from Resin Precursor)*

Basic properties:
Molecular weight... 3300–3500
Polydispersity index.. 1.58–1.66
Nitrogen, % (Kjeldahl).. 11.6–11.9
Moisture regain, %... 3.3
Melting point (DSC):
   First peak, °C.. 214–216
   Second peak, °C.. 219–221
Intrinsic viscosity (*m*-cresol) 0.25–0.29
Aqueous paste:
Solids, %... 44–45
Color.. White, opaque
Texture .... Smooth, no syneresis
Chlorides, %... <0.5
Ion conductance, grains of NaCl/gal................................. <4.0
pH.. 5.5–6.0
Electrophoretic mobility, $(\mu m/s)/(V/cm)$.......................... 1.9–2.1
Brookfield viscosity, cP (10 r/min, 10% solids)....................... 80,000–100,000
Suspendables, % (0.1% suspension test) (79).......................... 80–85 (less than 1 μm)
Powder form:
Appearance.. White, flowable
Moisture, % 0.5
Moisture, %
   (50% rh, 70°F ± 2°F).. 3.9
Bulk density, lb/ft^3... 28–31
Ash, % (ignition 700°C) 0.25
Extractions (H_2O and isopropanol), %............................. 2.5
Chlorides, %... 0.5
BET surface area, m^2/g... 6.6
Particle size, average in μm 30

Using their procedure, a 1 percent aqueous slurry of microcrystalline nylon-6 at pH 5.5 is allowed to stand for 24 h at $25 \pm 0.1°C$. The top fraction containing the suspendables is siphoned off the 1-liter graduated cylinder. The top and bottom fractions were freeze-dried. The dry-powder samples were stored in a desiccator over phosphorus pentoxide.

Utilizing the concept of the dependence of the falling velocities of particles on their size, equations were derived for the sedimentation techniques, where for a given particle diameter, particles above that diameter will sediment in a calculated time interval. The sedimentation techniques were verified by light and electron microscopy.

Differential-scanning-calorimetry Evaluation

Differential scanning calorimetry (DSC) is useful for the study of crystallinity in thermoplastic polymers. A Perkin–Elmer DSC-1 differential scanning

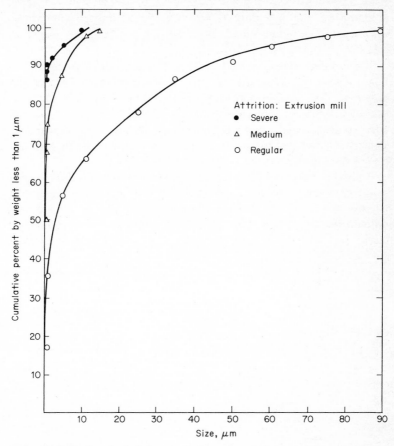

fig. 6.3 *Role of mechanical shear on deaggregation of microcrystalline nylon-6 microcrystals.*

calorimeter is suitable. A 10-mg sample of microcrystalline nylon-6 is heated in a nitrogen atmosphere at 10°C/min from room temperature to 237°C; a similar sample of microcrystalline nylon-66 is heated from room temperature to 277°C.

The first melts are cooled at 20°C/min to about 40°C. This cooling-down phase represents the *crystallization cycle*. The samples cooled in this way to about 40°C are now reheated over the appropriate temperature range to give the second-melt profile.

Figure 6.4 is a series of differential calorimetric scans of the precursor nylon-6 polymer. It shows a single endothermic peak during the first melt starting at about 213°C with peak development at 225°C. This broad endotherm is indicative of reduced crystallinity (less than 100 percent).

Figure 6.5 shows a parallel series of differential calorimetric scans for microcrystalline nylon-6 made from the precursor polymer shown in Figure 6.4.

The foregoing DSC scans show that the microcrystalline nylon-6 samples have melting temperatures about 5°C lower than the precursor nylon-6 polymer. The microcrystalline nylon-6 (residue after the controlled hydrolysis) also possesses exothermic peaks as compared with its polymer precursor, which reflects a higher total degree of crystallinity. One explanation of the somewhat lower melting point of the microcrystalline nylon form is that the size of the aggregated crystal particles is smaller.

It also is of particular interest to note the microcrystalline form exhibits sharp two-peak endotherms in both the first- *and* second-melt transitions, whereas the conventional precursor polymer shows a two-peak endotherm only during the second melt. This interesting behavior is shown with equal clarity for nylon-66 (Fig. 6.6) and microcrystalline nylon-66 produced from it (Fig. 6.7).

Nylon-66 has a single endotherm starting at 243°C with peak development

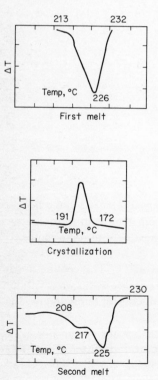

fig. 6.4 *Calorimetric scans of nylon-6 fiber (25).*

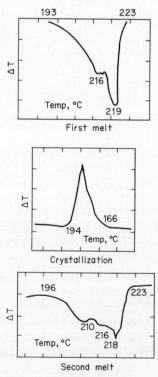

fig. 6.5 *Calorimetric scans of microcrystalline nylon-6, unfractionated (25).*

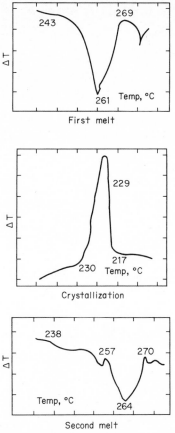

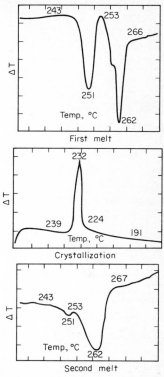

fig. 6.7   Calorimetric scans of microcrystalline nylon-66, bottom fraction (25).

fig. 6.6   Calorimetric scans of nylon-66 (25).

at 261°C. Microcrystalline nylon-66 shows a two-peak development at about 245°C and 260 to 263°C during the second melt.

Figure 6.8 demonstrates clearly the role of DSC in the characterization of the crystallinity of a polymer. It reveals the very broad endotherms and exotherms and much lower melting temperature found for the acid-soluble fraction recovered in the process of acid attack on the precursor nylon-6—the insoluble residue being the microcrystalline form. Such DSC scans are representative of highly amorphous polymer products.

Molecular Weight

The magnitude and nature of the molecular-weight distribution of the microcrystalline nylons were determined by gel-permeation chromatography (GPC). The GPC elution volume is correlated with the hydrodynamic

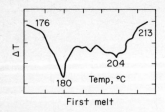

First melt

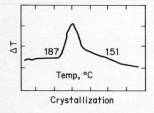

Crystallization

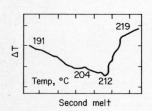

Second melt

fig. 6.8 Calorimetric scans of acid-soluble hydrolysate of microcrystalline nylon-6 (25).

volume of the polymer molecule in solution and not its extended-chain length. The possible complication of the ionic nature of the polymer must be considered. *m*-Cresol is used as a solvent, and the gel-permeation conditions are listed in Table 6.2. The GPC calibration curve utilized commercially available anionic polystyrene standards with molecular weights ranging from 5×10^3 to 2×10^6 (Fig. 6.9).

The elution tracings for microcrystalline nylon-6 reveal a narrow distribution of molecular weights. The average-molecular-weight range is from 3,300 to 3,500, and the polydispersity indices (ratio of weight-average molecular weight/number-average molecular weight) show a homogeneous polymolecularity. The molecular weight of the molecules in the microcrystals of nylon-6 is about one-seventh that of the precursor nylon.

Similar results are also obtained for samples of microcrystalline nylon-66. The microcrystalline polymer has an average molecular weight ranging from 3,900 to 4,100. This value is about one-twelfth that of the nylon-66 precursor (Table 6.2).

table 6.2   Molecular Weights of Microcrystalline Nylon-6 and Nylon-66 (25)

Type	$\bar{A}_w$	$\bar{A}_n$	$\bar{A}_w/\bar{A}_n$	$\bar{M}_w$	$\bar{M}_n$
Nylon-6:					
Microcrystalline	165.4	104.9	1.58	3300	2100
Microcrystalline, top fraction	177.9	107.5	1.65	3560	2150
Microcrystalline, bottom fraction	175.9	106.0	1.66	3250	2120
Acid-soluble hydrolysate	57.8	48.7	1.19	1160	975
Nylon-66:					
Microcrystalline	205.1	124.3	1.64	4100	2490
Microcrystalline, top fraction	201.8	113.4	1.78	4030	2270
Microcrystalline, bottom fraction	194.9	120.9	1.61	3900	2420
Acid-soluble hydrolysate	132.4	75.5	1.75	2650	1510

Notes:

$\bar{A}_w$ = weight-average molecular size in angstroms.
$\bar{A}_n$ = number-average molecular size in angstroms.
$\bar{A}_w/\bar{A}_n = \bar{M}_w/\bar{M}_n$ = polydispersity index.
$\bar{M}_w$ = weight-average molecular weight.
$\bar{M}_n$ = number-average molecular weight.

The hot-acid–soluble hydrolysate isolated in the preparation of the colloidal crystalline polymers has a molecular weight that is about half that of the microcrystalline material.

Viscosity and Rheological Properties

Viscosity-pH Dependence   Table 6.3 shows the dependence of apparent viscosities on pH and the reagent used to achieve a given pH level for microcrystalline nylon suspensoids.

Rheological Properties   A Haake Rotovisco instrument was used to study the structural recovery behavior of the microcrystal gel of nylon-6. The yield-stress buildup and the initial stress-decay behavior of aqueous nylon-6, cellulose, and regenerated-cellulose microcrystal gels are shown in Fig. 6.10. It is significant to note that microcrystals of regenerated cellulose and of nylon-6, each of which have almost the same dimensions (about 100 Å and nonrodlike in shape), exhibit similar yield-stress and stress-decay behavior in gel form. The smaller and more numerous microcrystals of regenerated-cellulose gels have a slow rate of yield-stress recovery compared to microcrystal gels prepared from wood-pulp cellulose. The differences in the rate of yield-stress recovery reflect the role of the size, shape, and particle-size distribution of the microcrystals recovered from their respective polymer precursors.

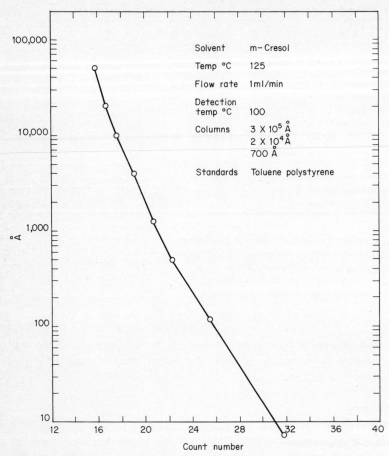

fig. 6.9 *Gel-permeation-chromatography calibration chart for molecular-weight measurements of microcrystalline polyamides (25).*

table 6.3 Relative Brookfield Viscosities of Microcrystalline Nylon Gels at 25°C

Reagent	pH	Solids, % (attrited in a Premier Colloid Mill)	Viscosity, cP
H_3PO_4	4.5	15	600
H_3PO_4	5.5	15	16,800
As is	6.8	15	1,520,000
NH_4OH	8.0	15	1,240,000
H_3PO_4	4.5	10	No reading
H_3PO_4	5.5	10	No reading
As is	6.5	10	64,000
NH_4OH	8.0	10	128,000

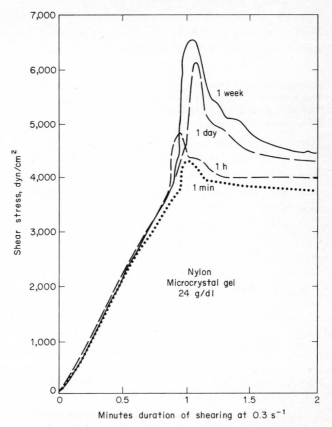

fig. 6.10   *Progression of yield-stress buildup after agitation of aqueous nylon-microcrystal gels (77).*

Some additional general properties observed for microcrystalline nylon-6 gels are:

1. Elevated temperatures as high as 85°C have no effect on the apparent viscosity of an 18 percent aqueous dispersion.

2. Freezing an aqueous suspension of microcrystalline nylon will cause separation of the particles to give a two-phase suspension; however, such suspensions may be readily redispersed and stabilized by mechanical means.

3. Acetic acid will not flocculate dilute microcrystalline nylon-6 dispersions.

4. Ammonium hydroxide causes flocculation of aqueous microcrystalline nylon suspensions when a pH of 7.8 to 8.0 is reached.

5. The surface tension of aqueous suspensions of microcrystalline nylon-6 is the same as that for water.

6. The smoothness and apparent viscosity of a microcrystalline nylon suspension increases sharply as the proportion of the individual discrete microcrystals in suspension increases.

Morphological Studies of Polyamides (Nylons) and Microcrystalline Polyamides Produced Therefrom

We have found that the basic individual microcrystals recovered from highly oriented nylon-6 textile fibers (Fig. 6.11) exhibit a discrete spherical size and shape of the order of 100 to 150 Å and that some of these are strung together somewhat like a string of beads that give the appearance of a helical submicron microfibril or microcrystal.

On the other hand, microcrystals recovered from a nylon-6 resin (from which nylon-6 textile fibers are made by melt spinning and stretching to produce a high degree of orientation), although they appear to have the identical basic microcrystal unit size, do not show the rodlike pseudobeaded linear submicron microfibrils (Fig. 6.12). This is even more clearly evident in Fig. 6.13, which shows the gross shape and much bigger size of nylon-12 particles produced by extensive mechanical grinding of commercial nylon-12 resin.

When polycaprolactams of low molecular weights are produced by catalytic polymerization of the monomer, plate-like particles result which consist of side-by-side–packed microcrystals of the same fundamental unit size (100 to 150 Å) one layer thick to give flat structures (Fig. 6.14).

It would seem significant that the unit microcrystal (about 100 Å) so clearly isolated from nylon-6 textile fibers (Fig. 6.11) is the identical building block

1 μm

fig. 6.11   *Nylon-6 microcrystals from yarn.*

fig. 6.12 *Nylon-6 microcrystals from nylon pellets.*

from which the particle aggregates are built in nylon-6 granular resins or in aqueous catalyzed polycaprolactams—polycaprolactams polymerized up to low-molecular-weight levels consistent with microcrystalline nylons.

Microcrystalline nylon-6 exhibits a very sharp x-ray diffraction pattern indicating a very high degree of crystallinity (Fig. 6.15).   Figure 6.16 illus-

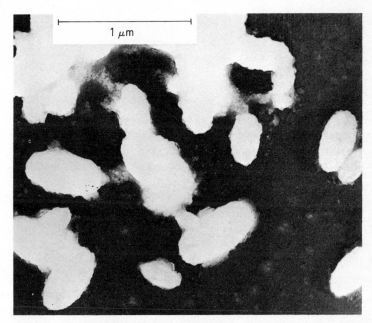

fig. 6.13 *Commercial ultrafine nylon-12 powder.*

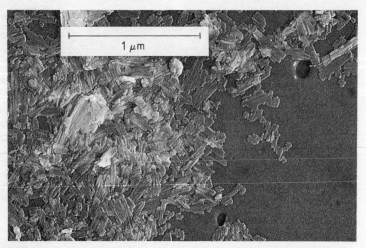

fig. 6.14　*Nylon-6 platelet microcrystals from the catalytic polymerization of caprolactam.*

trates the pseudoplastic rheology of aqueous microcrystalline nylon suspensoids, and Fig. 6.17 shows the electrophoretic mobility of microcrystalline nylon-6 with changing pH.

A most interesting phenomenon has been the finding that the electrical conductance of microcrystalline nylon-6 suspensoids is clearly related to the proportion of free microcrystals or particles in the far-submicron size range (Fig. 6.18).

APPLICATIONS

Conventional polyamides are relatively insensitive to water, but microcrystalline polyamide products, particularly those having essentially all particles below 1 μm in size and containing a high proportion below 0.01 μm in size, are dispersible in water to form stable dispersions and gels. Other satisfactory media to produce stable suspensoids include dilute aqueous solutions of lower aliphatic acids such as formic, acetic, dichloroacetic, and trichloroacetic acids; lower aliphatic alcohols such as methanol, ethanol, and isopropanol; mixtures of phenol, cresol, and resorcinol with water; ketones such as acetone; aldehydes such as formaldehyde and acetaldehyde; and other organic solvents such as formamide, dimethylformamide, etc.

Initial market development of microcrystalline nylons shows promise for them in the glass fibers or cloth industry. The nylon microcrystals may be applied to such systems by dipping the substrate into an aqueous gel dispersion of concentrations of the order of 5 to 10 percent. The water can be easily removed by drying at 105°C to provide a uniform protective coating

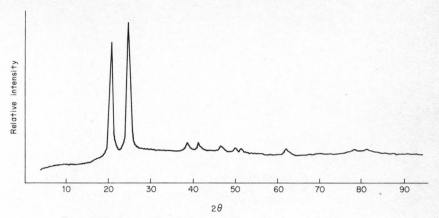

fig. 6.15 *X-ray diffraction pattern of microcrystalline nylon-6.*

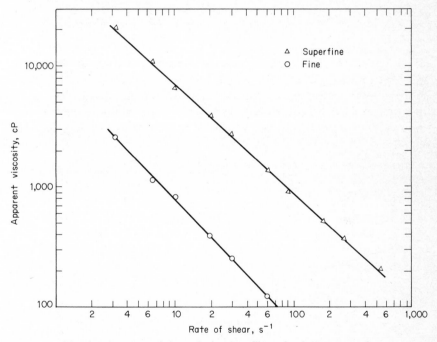

fig. 6.16 *Viscosity versus rate of shear of microcrystalline nylon-6 15 percent gels.*

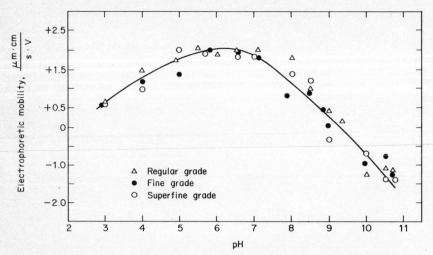

fig. 6.17   *Electrophoretic mobility versus pH of microcrystalline nylon-6 15 percent gels.*

on the glass.   Treated fibers are improved in terms of strength, dyeability, and fiber friction.   In the case of glass fabrics for drapery and industrial uses, the nylon microcrystals can be applied as a sizing agent to reduce fiber-to-fiber friction and also to improve fabric hand, dullness, and stability. As a permanent anchoring agent on glass-fiber surfaces, coatings improve

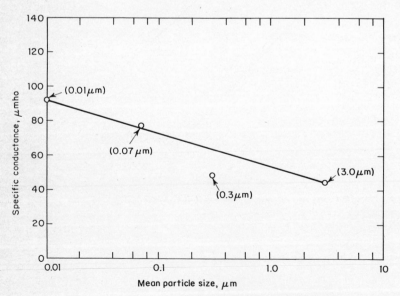

fig. 6.18   *Increase of electrical conductance with reduction in size of nylon microcrystal aggregation.*

the bond between glass and plastics and/or rubber composites such as tires. Figure 6.19 illustrates the use of microcrystalline nylon as a glass-cloth adhesive or as a glass-fiber coating to improve the dyeing characteristics of glass textiles.

Microcrystalline nylons have a demonstrated potential for use in paints and pigment dispersions generally. They permit a reduction in the titanium dioxide content of some latex-type paints such as those based on polyvinyl acetate emulsions. Simultaneously, they make possible an increase in the calcium carbonate content without noticeably affecting the covering power or brilliance of the paint. The presence of the microcrystalline polyamide also contributes to the latex-particle stability of the system. Obviously, any desired coloring material may be added to provide a desired tint or color to the finished paint.

The polyamide products are also useful as gellants and carriers for coloring materials in inks. The products add to the stability of the inks and can be used to impart thixotropic properties. They may replace various gums and resinlike materials conventionally employed in various inks.

As a protective coating, dye absorptive, and permanent-press agent, the nylon microcrystals can be applied as an aqueous gel as a warp sizer and dried. When woven fabrics prepared from the coated glass yarns are dried in the neighborhood of 160 to 170°C, the fabrics are stabilized structurally without coronization (burning off) or loss of the protective coating. Permanent-press properties are achieved with fabrics made from textiles other than glass fibers by the application of appropriate heat and pressure.

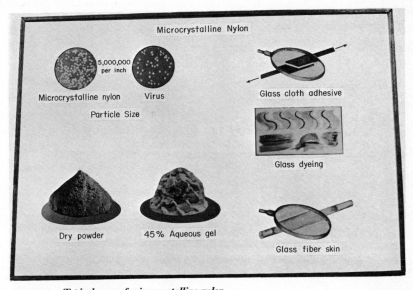

fig. 6.19 *Typical uses of microcrystalline nylon.*

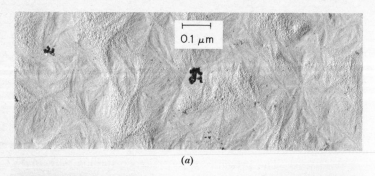

(a)

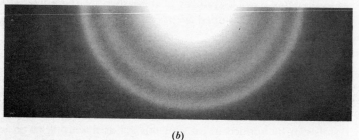

(b)

fig. 6.20 (a) *Ultrathin film of microcrystalline nylon-6 on aluminum substrate* (*from aqueous gel*); (b) *electron diffraction pattern.*

Other applications that have been examined for microcrystalline nylons are the preparation of membranes with controlled porosity, such as an electrographic binder for photoconductive papers, a dispersant for carbon black in lithographic offset processes, and a binder for carbon black in carbon papers.

Shaped articles may be formed from stable microcrystalline nylon dispersions by extruding or casting the dispersion in the desired shape, and then washing or immersing the article in water or a dilute alkaline solution such as sodium hydroxide; or the microcrystalline polyamide gel or dried attrited material may be blended with an appropriate plasticizing agent and the blends extruded, molded or cast into various shapes, and dried. The dispersions and gels are also useful in the production of films, as coatings for various base materials such as, for example, aluminum, paper, wood, etc., and as a binding agent in nonwoven fabrics.

Figure 6.20a shows an ultrathin coating of microcrystalline nylon applied to an aluminum surface from an aqueous suspension, dried, and fused. An electron diffraction diagram of the microcrystalline nylon used for this coating comprises the bottom half of the illustration (Fig. 6.20b).

CHAPTER SEVEN

Microcrystalline Polyesters

INTRODUCTION

Polyethylene terephthalate (PET) is the most abundant polyester polymer for use in fibers and films. The raw materials from which it is produced are ethylene glycol (EG, bp 197°C), which is obtained from ethylene via ethylene oxide, and terephthalic acid (TA, mp circa 300°C, with sublimation), which is obtained by oxidation of *p*-xylene with nitric acid or directly with air.

The resulting polyester polymers possess unusually good acid and alkali resistance, and their morphology comprises crystalline regions in a viscous network.

Most commercial processes use the following steps to produce the polymer:

1. *Esterification* of methanol with TA to yield dimethyl terephthalate (DMT, mp circa 140°C when purified—as is readily done—by recrystallization).

2. *Ester interchange* between DMT and an excess of EG at 150°C, to yield largely bis 2-hydroxyethyl terephthalate (with recovery of methanol by distillation),

$$CH_3O\underset{O}{C}-\underset{}{\bigcirc}-\underset{O}{C}OCH_3 + 2HO(CH_2)_2OH \longrightarrow$$

$$HO(CH_2)_2O\underset{O}{C}-\underset{}{\bigcirc}-\underset{O}{C}O(CH_2)_2OH + 2CH_3OH$$

3. *Polycondensation* of the above product, under vacuum at 275°C (with continuous removal—and recovery—of EG),

$$HO(CH_2)_2OC{-}\langle\rangle{-}CO(CH_2)_2OH \longrightarrow$$
$$\qquad\qquad\; \overset{\|}{O}\qquad\qquad \overset{\|}{O}$$

$$H\left[-O(CH_2)_2OC{-}\langle\rangle{-}C-\right]_n O(CH_2)_2OH + (n-1)HO(CH_2)_2OH$$

When the final melt reaches the required viscosity, the polymer is rapidly cooled to avoid degradation and discoloration; for extrusion as fibers or films it is subsequently remelted under nitrogen.

Polycarbonates are a form of polyesters. They are transparent faintly amber-colored thermoplastic materials showing good dimensional stability, thermal resistance, electrical properties, and tensile and impact strength which are suitable for moldings, fibers, and films.

The monomer of the commonest polycarbonate, based on diphenylolpropane, is represented by

$$-\langle\rangle{-}\overset{\overset{\displaystyle CH_3}{|}}{\underset{\underset{\displaystyle CH_3}{|}}{C}}{-}\langle\rangle{-}OCO-$$
$$\qquad\qquad\qquad\qquad\qquad\qquad \overset{\|}{O}$$

$C_{16}H_{14}O_3$, mol. wt. = 254

PREPARATION

Microcrystalline polyesters that result in unusually smooth aqueous suspensoids are prepared by mechanical comminution of a partially hydrolyzed polyester. The polyester is subjected to (1) hydrolysis under conditions such that only the amorphous parts of the resin are hydrolyzed, which is followed by (2) mechanical abrasion or rubbing of the microcrystalline aggregates until at least about 5 percent of the colloidal particles are well under 1 μm, preferably 50 percent or more are under 1 μm.

Such finely divided polyester particles form stable dispersions in liquids which have a minimum swelling effect on the microcrystals and which can solvate the surface without attacking the interior of the microcrystals. The dispersions can be used for extrusion or casting or as coating compositions, or they can be dried.

It is known that synthetic linear polyesters possess a crystalline-amorphous network, or a morphological structure consisting of regions of high order commonly referred to as crystalline and regions of low order commonly referred to as amorphous. In this network structure, primary chains extend through a series of microcrystals connecting them by amorphous regions or hinges involving primary molecular bonds. Because of the strength of the primary molecular bonds it is impractical to free the microcrystals by mechanical means, such as attrition or grinding. Dissolving and reprecipitation of the polyester results merely in a rearrangement of the crystalline amorphous network, which leads once more to a continuous network of crystalline and amorphous areas connected by primary molecular bonds.

It is also known that synthetic linear polyesters having molecular weights sufficiently high to allow conversion into fibers possess toughness and elastic characteristics which make grinding of the polyesters into finely divided particles exceedingly difficult. Precipitation of such polyesters from a solution tends to produce stringy, cohesive masses. While low-molecular-weight polyesters can be ground to a finely divided state, the resulting particles are still characterized by a typical polyester crystalline-amorphous network. They do not form stable colloidal dispersions and gels in water.

Laboratory conditions have been developed to prepare microcrystalline polyesters at high yields. Weight-average molecular weight values of about 10,000 have been measured for microcrystalline polyester products using gel-permeation-chromatography techniques; the precursor polymer from which such microcrystalline polyesters are produced has a molecular weight of about 40,000. The x-ray diffraction pattern of the microcrystalline polyester residue invariably shows a marked increase in sharpness of crystallinity peaks over the precursor polymer.

The choice of the hydrolytic agent as well as the swelling medium is quite critical to achieve maximum yields of a well-defined microcrystalline polyester product, and these choices are dependent upon the particular class of polyester precursor chosen.

The hydrolytic agent may be any of the types capable of cleaving ester linkages, although basic materials are preferred. Suitable bases include both the organic and inorganic members, and in this connection reference is made to the lower aliphatic amines, e.g., ethylamine, n-propylamine, urea, and ammonium hydroxide and its alkylated derivatives, as well as such inorganic bases as the alkali-metal carbonates and hydroxides. An especially effective hydrolytic medium, particularly for polycarbonate resins, consists of ethylacetate containing a small quantity of ethanolic sodium hydroxide. When polycarbonate resin granules or fibers are refluxed for a few hours in the previously described mixture, they are converted almost quantitatively into the microcrystalline morphological form.

A characteristic of microcrystalline polymers which have been substantially freed of *amorphous* areas is the ease with which they can be crushed to a fine powder. In fact, the process of hydrolysis can be conveniently followed in a suitable manner for optimizing the choice of the hydrolytic medium as well as monitoring the course of the reaction. All that is required is that samples of the particular polyester resin be removed from the reaction mixture from time to time, dried, and placed between two microscope slides and that the slides be gently rubbed together with mild finger pressure. Once the amorphous areas have been loosened leaving a matrix of microcrystalline aggregates, the polymer particles are readily crushed when rubbed between the glass slides.

If the polymer particles become sticky or tacky, this signifies that the solvent medium is too strong and that the crystalline network is being unduly swollen, or even partially dissolved. On the other hand, if after lengthy hydrolytic treatment no sign of friability between the glass slides has developed, this signifies that the particular solvent medium failed to cause even minimal swelling of the polymer so as to enable the hydrolytic agent to penetrate to the amorphous areas.

So far as we have been able to ascertain, the point of attack of the hydrolytic reagent is directed toward the ester linkage, and as a consequence, the terminal groups in the hydrolyzed polymer will be carboxy and hydroxy functions when using mineral bases, whereas the use of amines will form a terminal amide group.

Although the polyester precursor (fiber or resin) may be used directly as purchased or prepared, the polyester is preferably subjected to a preannealing treatment in order to increase its crystallinity. A typical procedure consists of heating polyester resin pellets at about 150°C for 1 h in a nitrogen atmosphere. Another satisfactory annealing procedure consists of immersing the pellets in an inert liquid such as Dow-Corning Silicon 200 for 4 h in the general range of 125 to 150°C. The treated material is washed free of silicon oil with a suitable solvent such as acetone and then air-dried.

The isolated microcrystalline polyester product in aggregated form is converted into free microcrystals by mechanical shearing action of the type provided by attrition mills, planetary mixers, sonic mixers, grinding mills, and the like. The mechanical breakdown of the partially degraded polymer is effectively accomplished in a liquid which exerts a controlled, limited swelling effect on the microcrystals, and for the purposes of this invention such a liquid can be termed a *liquid swelling medium.* This liquid medium has a minimum swelling effect on the microcrystals and may solvate or tend to solvate the surface of the individual microcrystals but does not attack and destroy the lateral order or crystallinity in the interior of the microcrystals. Although the polyesters are relatively insensitive to water, the microcrystalline polyester products of this invention, particularly those having essentially all material below 1 μm in size and containing particles below 0.01 μm in size, are dispersible in water to form stable dispersions and gels. Water is therefore

fig. 7.1 Aqueous gel of microcrystalline polyester (20 percent).

deemed a satisfactory liquid swelling medium. Other satisfactory swelling agents include, for example, dilute aqueous solutions of lower aliphatic acids such as formic, acetic, dichloroacetic, and trichloroacetic acids; lower aliphatic alcohols such as methanol, ethanol, and isopropanol; mixtures of phenol, cresol, and resorcinol with water; ketones such as acetone; aldehydes such as formaldehyde and acetaldehyde; and other organic solvents such as formamide, dimethylformamide, etc.

The following procedure is an example of how polyester fiber (PET) is converted into the microcrystalline state.

A mixture of 10.0 g of polyethylene terephthalate fiber, 3.5 g of *n*-propylamine, 1.75 ml of ethylene glycol, and 134.5 ml of water is heated without agitation at 150°C ± 5°C for a period of 3 h. The reaction product is washed with water until the pH of the washings is about 7 to 8. The yield of microcrystalline product was 80 percent. The product is subsequently air-dried or vacuum-dried to a fine white powder. Stable, smooth aqueous suspensoids are readily made from it (Fig. 7.1). PET microcrystals from this fiber precursor are shown in Fig. 7.2. Note the typical spherical microcrystals

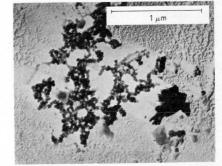

fig. 7.2 Microcrystals from commercial PET (Dacron) fibers.

recoverable from highly oriented PET fibers, which parallels particles recovered from highly oriented nylon fibers (Fig. 6.11).

Microcrystalline Polycarbonate (Polyester)

It is possible to convert a polycarbonate (polyester) such as bisphenol A by the following procedure:

Starting Composition

60.0 g bisphenol A polycarbonate

600.0 ml ethyl acetate

30.0 ml solution prepared by adding 5 ml of a solution containing 1.0 g of sodium hydroxide per 99.0 ml of absolute ethanol to 100 ml of ethyl acetate.

Reaction Conditions   The above composition is refluxed for 3 h, cooled and washed with ethyl acetate, partially dried, and then washed in a Waring Blendor for 1 to 2 min with water.   The resulting powder is then dried and fluffed for 1 min in a Waring Blendor.

Gel Formation   The preparation of a stable, nonsettling, thixotropic microcrystal gel is achieved by taking 25 g of the above product and attriting it in a Waring Blendor for 1 h in a ternary medium of 70 percent isopropanol/20 percent water/10 percent CCl_4.

The resulting gel gives an apparent viscosity of 38,000 cP in equilibrium at 25 wt percent concentration.   Electron micrographs (Fig. 7.3) reveal the presence of 0.1-μm-size platelets.

The x-ray diffraction spectrum (Fig. 7.4) shows the high level crystallinity of microcrystalline polyester prepared from a fibrous PET Dacron-type precursor.

Investigations were carried out to determine the reaction mechanism of the breakdown into microcrystals.   Assuming that it is essentially a hydrolysis mechanism, the NaOH was replaced with an equivalent amount of toluene sulfonic acid, itself a catalyst for hydrolysis in the above reaction mixture. No breakdown of structure occurred.   Hence it is concluded that the reaction is most probably not a result of a nucleophilic attack of the OH^- ion but rather of the stronger $OC_2H_5^-$ ion on the carbonyl group, which results on the one hand in an ethyl half-ester of the polymeric carbonic acid and on the other hand in an Na^+ salt of the polymeric bisphenol A, whose two ends have previously formed linkages of the polycarbonate with other molecules. However, we cannot rule out the possibility of an ester interchange between the polymeric resin and the monomeric solvent as a side reaction, which is indicated by the partial friability of the polycarbonate resin when refluxed in ethyl acetate alone.

fig. 7.3   *Microcrystalline polycarbonate.*

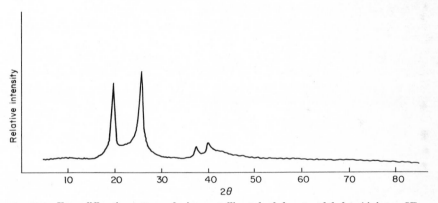

fig. 7.4   *X-ray diffraction pattern of microcrystalline polyethylene terephthalate (Aviester, SF).*

PROPERTIES

Microcrystalline polyesters (from PET) possess two distinct endotherms, 227 and 245°C, as evidenced by differential calorimetric scans.   In line with findings about earlier members of the microcrystalline polymer family, the polyester microcrystals possess a much higher degree of lateral order (crystallinity) than does the starting resin.

Unlike their polymer precursor, they possess an amide end group.   However, the presence of this group does not appear to exert a pronounced

table 7.1

Polymer type	Precursor PET polymer	Microcrystalline polyester*
Molecular weight, $\overline{M}_w$	60,000	3,000–4,000†
Density, g/cm^3	1.342	1.410–1.413
Intrinsic viscosity, $[\eta]$‡, 30°C	0.619	0.080–0.100
N, %	0.0	0.35–0.40
Melting point (DSC):		
First peak, °C	None	230–236
Second peak, °C	250–255	244–247
% crystallinity§	5	66

*Ranges given to cover regular and superfine grades.
†Estimated from % N in absence of GPC results.
‡60% phenol/40% s-tetrachloroethane.
§Calculated from density in lieu of x-ray diffraction results.

influence on their properties other than to make the product in the colloidal state more compatible in aqueous suspension.

The preceding table provides data comparing the PET precursor polymer with the microcrystalline form derived from it.

Microcrystalline polyester (PET) has been evaluated in two grades. Grade R is a regular grade and Grade SF is a superfine grade. The difference in the two grades is a reflection of the proportion of superfine particles built up in the suspension which is the result of a difference in the degree of mechanical disintegration of each. Typically the suspensoids exist as aqueous gels with a range of solids between 20 and 50 percent. These suspensoids are pseudoplastic and exhibit thixotropic behavior.

table 7.2 **Apparent Viscosity (cP $\times$ 10^3)** *·† **of Aqueous Microcrystalline Polyester Suspensoids**

Regular grade‡	Shear rate, r/min	Superfine grade§
140.00	0.5	264.00
94.60	1.0	150.00
44.00	2.5	65.90
24.80	5.0	34.60
10.40	10.0	18.30
4.80	20.0	9.60
1.72	50.0	4.37
0.80	100.0	2.33

*At 25°C.
†Brookfield viscosity, after 48 h aging.
‡31% solids.
§20% solids.

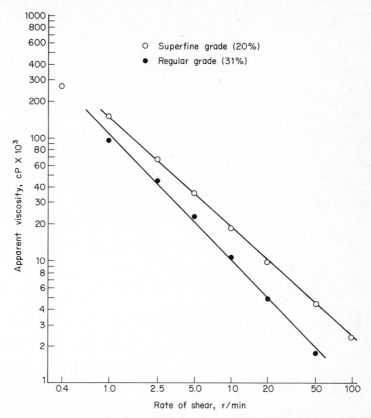

fig. 7.5 *Apparent viscosity at 25°C for microcrystalline polyester gels.*

Apparent viscosities of these gels as measured with a Brookfield viscometer are given in Table 7.2 and plotted in a log-log presentation in Fig. 7.5. These gels show a tendency to age on standing, yielding higher viscosities with time. It would appear that solvation of the particles continues on aging.

The natural pH range of these suspensions in distilled water is 8.0 to 8.5. They are of low conductivity (5 to 10 parts per million as NaCl). The colloidal particle is negatively charged (anionic) and is exceptionally stable without coagulation between the pH's of 6 and 10, using NH_3 and HOAc to change the hydrogen-ion concentration. Varying the pH outside this range causes complete precipitation of the coagulated colloid. This behavior results from a neutralization of the negatively charged colloid surface by cations such as NA^+, H^+, or Ca^{++}. Hence, adjustment of the pH within the 6 to 10 range should be made with the weak electrolytes suggested above. These observations are corroborated by zeta-potential measurements made on a 20 percent aqueous SF microcrystalline polyester slurry as shown in

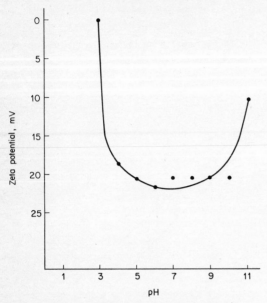

fig. 7.6 *Microcrystalline polyester SF-zeta-potential measurements as a function of pH (20 percent gel).*

Fig. 7.6. It would appear that the pH range of maximum stability coincides with the minimums in this figure. Outside this range, the charge and mobility of the colloid are reduced.

The addition of strong caustic (1 percent of NaOH or greater) causes coagulation of the microcrystalline polyester, followed by a gradual degradation into a water-soluble product (disodium terephthalate). The rate of this reaction depends on the concentration of the alkali present.

Additional data comparing properties of the regular and superfine grades of microcrystalline polyesters are provided in Table 7.3.

Morphological Studies of Polyesters and Microcrystalline Polyesters Produced Therefrom

Paralleling the observations made for nylon-6 fibers and resins, polyester microcrystals recovered from a precursor polyethylene terephthalate granular resin (precursor for the fiber) show a more platelet-like structure and evidence of the unit microcrystals growing into flat wafers only one microcrystal thick in the unaggregated state (Fig. 7.7). This electron micrograph should be compared with Fig. 7.2, showing microcrystals recovered from highly oriented PET fibers.

Regular grade	Property	Superfine grade
3000–4000	Molecular weight, $\overline{M}_w$*,†	3000–4000
$\approx$0.4	Nitrogen content, %	$\approx$0.4
1.411	Density, g/cm³	1.413
0.093	Intrinsic viscosity, $[\eta]$‡, 30°C	0.09–0.11
0.1	Ash, %	0.2
0.9	Moisture, %	1.5
White	Color	White
150 mesh	Particle size	150 mesh
	Melting point (DSC):	
234–236	First peak, °C	230–232
244–246	Second peak, °C	244–247
	Aqueous gel dispersions:	
8.0–8.5	pH	8.0–8.5
—	Electrophoretic mobility, $(\mu m)(cm)/(V)(s)$	−1.60
10.4 × 10³	Apparent viscosity§, cP (HAT, 10 r/min 25°C)	18.3 × 10³
	Particle size:	
62	$\lesssim$1.0 μm	80
39	$\lesssim$0.2 μm	70

*Tentative.

†Estimated from % N content in absence of GPC data.

‡60% phenol/40% s-tetrachloroethane.

§Data for 31% (wt./wt.) microcrystalline polyester-R and 20% (wt./wt.) microcrystalline polyester-SF in water.

fig. 7.7 *Microcrystals recovered from polyethylene terephthalate resin.*

fig. 7.8 *Polyester (PET) microcrystals from PET granular resin precursor (60 min at high speed in an Osterizer, supernatant particles diluted to 0.1 percent).*

Figure 7.8 is an electron micrograph of microcrystals recovered from the same polyester resin precursor as that for Figure 7.7 except that the following more severe mechanical attrition was used to free the constituent microcrystals in their smallest unit entities:

A 10 percent solids suspension of the microcrystalline polyester was mixed at high speed in an Osterizer blender for 60 min. The electron micrograph was obtained from an aliquot of the supernatant of a 0.1 percent suspension after being allowed to stand 16 h.

APPLICATIONS

Inasmuch as extremely stable aqueous suspensoids of extremely small microcrystals of polyethylene terephthalate can be made, they offer interesting possibilities for many uses. The acid and alkali resistance of saturated polyester polymers is excellent, the polymers possess high temperature stability, and the gas and vapor permeability of coatings and/or films made from them are known to have proven commercial value.

Shaped particles may be formed from the stable aqueous dispersions, for example, by extruding or casting the dispersion in the desired shape and washing or immersing the article in water or a dilute alkaline solution, such as sodium hydroxide. Or the microcrystalline polyester gels or the dry attrited material may be blended with an appropriate plasticizing agent, and the blend extruded, molded, or cast into various shapes and dried. The dispersion and gels are useful in the production of films; as coatings for various base materials, such as paper, wood, and the like; and as binding agents in nonwoven fabrics. The microcrystalline polyester suspensoids are also useful for admixture with other microcrystalline products such as microcrystalline cellulose, microcrystalline starch, or microcrystalline nylon in the production of molding powders and structural objects. The spray-dried gels give ultrafine polyester powders that are suitable for fluidized-bed powder coating applications.

CHAPTER EIGHT

Microcrystalline Polyolefins

INTRODUCTION

Essentially all of our studies of microcrystalline polyolefins have been with highly crystalline (isotactic) polypropylene fibers and/or resins or pellets.

Once again the approach has been to subject the precursor polymer to a controlled chemical attack, in the present instance a thermal catalytic oxidation whereby molecules in the limitedly accessible areas of the polymer-precursor morphology are cleaved, degradation fragments are solubilized and removed, and the more crystalline, difficult-to-oxidize residues remain.

Subsequent mechanical attrition of the microcrystalline polyolefin residue is preferably performed in the presence of a liquid having a minimal swelling action on the crystalline polyolefin, which in some instances may solvate some of the surface thereof but which will not destroy its particulate or discrete form. Mechanical attrition is continued until a stable colloidal dispersion results. In many instances, when at least about 5 wt percent of the polyolefin is reduced to a particle size not exceeding 1 μm, the stable dispersion will be attained. Preferably, however, attrition is continued until at least about 95 wt percent of the polyolefin does not exceed a particle size of 1 μm and at least about 10 wt percent has a particle size not exceeding 0.1 μm. Some of the disintegrated polyolefin microcrystals will have a size under 0.01 μm.

PREPARATION

Preparation of a microcrystalline polyolefin suspension having the appearance and properties paralleling other aqueous microcrystalline suspensoids such as microcrystalline celluloses, microcrystalline polyamides, and microcrystalline polyesters is best done using the following procedure.

It is essential to carry out the oxidation of the precursor isotactic polypropylene with oxygen in a medium where a substantial amount of CCl_4 is present (30 percent vol) with the heptane and acts synergistically with the copper stearate catalyst. Parallel tests showed that in the presence of CCl_4 the breakdown is about an order of magnitude faster than in heptane alone, it is more thorough, and we could successfully gel only such a product as was oxidized in a mixed CCl_4-heptane medium. The yield obtained was 90 to 93 wt percent, and x-ray diffraction analysis showed increased crystallinity, while DSC analysis indicated slightly lowered melting point and smaller particles by comparison with the starting material.

Preparation of a Stable Colloidal Suspensoid of Microcrystalline Polypropylene

Reaction Mixture
20 g polypropylene as coarse resin powder
320 ml solvent mixture whose volume ratio is 30 percent CCl_4/70 percent heptane
0.6 g copper stearate
80 lb/in^2 oxygen pressure

Conditions of Reaction The precursor isotactic polypropylene is oxidized for 2 h at 100°C in oil bath using a stainless-steel pressure bomb with a gauge with rapid pendulum-like shaking, where stirring is achieved by inserting a 1-in-diameter glass ball (marble).

At the end of 2 h the pressure bomb is placed into cold water to cool, pressure is released, and the bomb is opened.

Removal of Catalyst and Impurities The contents of the bomb are poured onto a Buchner funnel, rinsed and washed with 200 ml isopropanol, and then filtered.

The product next is reslurried in a Waring Blendor for 1 to 2 min at low speed in a 250 ml mixture whose volume proportions are 1 part isopropanol/2 parts concentrated HCl/3 parts H_2O. The presence of isopropanol is necessary for wetting, while the acid dissolves the various copper compounds, which are removed down to 10 parts per million concentration or less.

Product is next filtered on Buchner funnel.

Conversion of Microcrystalline Polypropylene to Dry Powder Form The wet cake is put into a Waring Blendor, 500 ml of H_2O are added, and it is attrited at low speed for 1 to 2 min. This washing removes the acid and fluffs the powdery polypropylene, which is filtered again and washed with more water.

Next, the loose powdery cake is dried in an oven at 70°C. The dried microcrystalline polypropylene is a smooth, fluffy powder which is readily broken down into microcrystals on shearing or crushing.

X-ray diffraction of this product reveals an increase in crystallinity, while DSC analysis shows the presence of a reduced unit-crystal size.

Gel Preparation from Microcrystalline Polypropylene   Twenty grams of microcrystalline polypropylene is added to 80 ml of a blend whose volume ratio is 70 percent isopropanol/20 percent H_2O/10 percent CCl_4 and well mixed.

This composition gives 22 wt percent solids and is too thick to be stirred effectively in a Waring Blendor. Additional liquid must be added, for example, 43 ml of the above ternary mixture, which reduces the concentration to 15 wt percent.

PROPERTIES

Catalytic oxidative degradation of isotactic polypropylene and its attendant conversion to the microcrystalline polymer state has, as would be expected, a pronounced effect on molecular weight. This change is shown in Table 8.1. The effect on molecular-weight distribution is shown in Table 8.2.

The number-average and weight-average molecular weights of both samples were determined by means of gel permeation chromatography.

The polydispersity or molecular-weight distribution of polypropylene is changed markedly upon converting it into microcrystalline polypropylene; a noticeable increase in the uniformity of the polydispersity occurs during the transition.

Conversion of polypropylene to the microcrystalline polymer form leads to a very definitive increase in density, crystallinity, and surface area (Table 8.3 and Fig. 8.1).

table 8.1   Molecular Weight of Microcrystalline Polypropylene

Material	$[\eta]$	$\overline{M}_w$	$\overline{M}_n$	$\overline{M}_w/\overline{M}_n$
Polypropylene (powder)	2.62	433,700	148,600	2.92
Microcrystalline polypropylene (powder)	0.30	34,770	21,000	1.66

Notes:

$[\eta]$ = Intrinsic viscosity.
$\overline{M}_w$ = Weight-average molecular weight.
$\overline{M}_n$ = Number-average molecular weight.
$\overline{M}_w/\overline{M}_n$ = Polydispersity index.

table 8.2 *Molecular Weight Distribution of Microcrystalline Polypropylene*

Polypropylene powder*		Microcrystalline polypropylene	
Molecular weight, range	Material, %	Molecular weight, range	Material, %
$0\text{--}50 \times 10^3$	7.00	$0\text{--}\ 5 \times 10^3$	1.19
$50 \times 10^3\text{--}100 \times 10^3$	12.70	$5 \times 10^3\text{--}\ 10 \times 10^3$	7.96
$100 \times 10^3\text{--}150 \times 10^3$	11.10	$10 \times 10^3\text{--}\ 20 \times 10^3$	22.50
$150 \times 10^3\text{--}200 \times 10^3$	9.40	$20 \times 10^3\text{--}\ 30 \times 10^3$	21.20
$200 \times 10^3\text{--}500 \times 10^3$	31.50	$30 \times 10^3\text{--}\ 40 \times 10^3$	15.50
$500 \times 10^3\text{--}\ 1 \times 10^6$	17.40	$40 \times 10^3\text{--}\ 50 \times 10^3$	10.80
$>1 \times 10^6$	10.90	$50 \times 10^3\text{--}\ 60 \times 10^3$	7.00
		$60 \times 10^3\text{--}\ 75 \times 10^3$	6.40
		$75 \times 10^3\text{--}100 \times 10^3$	5.00
		$>100 \times 10^3$	2.50

*Starting raw material.

table 8.3 *Changes in Basic Physical Properties of Polypropylene on Conversion to Microcrystalline Polypropylene*

Treatment	Density, g/cm³	X-ray crystallinity, %	Surface area, m²/g
Precursor polypropylene	.890-.894	47.8	5.4
Microcrystalline polypropylene	.922	72.1	38.0

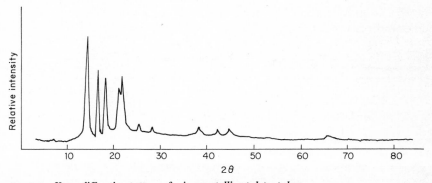

fig. 8.1 *X-ray diffraction pattern of microcrystalline polypropylene.*

The x-ray crystallinity data were determined by measurement of the integrated areas of the crystalline reflections and the noncrystalline background. The experimental accuracy is within ± 2 percent.

Change in Refractive Index on Conversion to Microcrystalline Polypropylene

As the data in Table 8.4 show, there is a significant change in the refractive index upon converting polypropylene in either fiber, granular, or powder forms into the corresponding microcrystalline end products.

Microcrystalline polypropylene exhibits double endotherms not present in the precursor polymer when evaluated by differential calorimetric scanning. In this respect, its properties closely parallel those observed and already commented upon for microcrystalline polyamides and microcrystalline polyesters, respectively. Figures 8.2 and 8.3 compare the differential calorimetric scans for both the precursor polypropylene and the microcrystalline polymer product produced from it.

The Perkin-Elmer DSC-1B differential scanning calorimeter was used to evaluate the melt behavior of the microcrystalline polypropylenes. A 5-mg sample is heated in a nitrogen atmosphere at a rate of 10°C/min from room temperature to about 187°C. This constitutes the first melt. The melt is cooled at 10°C/min to below 50°C. This represented the crystallization cycle. The cooled sample is reheated under the same conditions as the first melt. This is designated as the second melt (25).

A convenient reference summary of physical properties of microcrystalline polypropylene is compiled in Table 8.5.

Figure 8.4 provides a direct comparison of the infrared spectra of polypropylene and microcrystalline polypropylene.

The infrared spectra of polypropylene and the microcrystalline polypropylenes were determined with a Perkin-Elmer Model 137 infrared spectrophotometer. The powder sample is intermixed intimately with Nujol between two glass slides. The thick slurry is mounted on the NaCl sample disks and placed in the cell.

table 8.4 Changes in Refractive Index—Microcrystalline Polypropylene

Treatment	Form	Refractive index Vertical	Horizontal
Precursor polypropylene	Fiber	1.492	1.512
Microcrystalline polypropylene	Readily friable fibers	1.496	1.528
Precursor polypropylene	60- to 100-mesh granules	1.500	
Microcrystalline polypropylene	Fine powder	1.512	

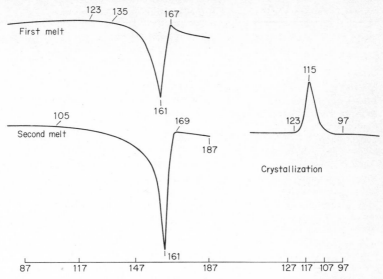

fig. 8.2 *Differential calorimetric scan of polypropylene powder (control) (25).*

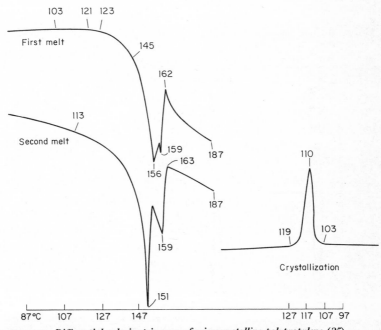

fig. 8.3 *Differential calorimetric scan of microcrystalline polypropylene (25).*

table 8.5   Summary of Properties of Microcrystalline Polypropylene

	Polypropylene (precursor)	Microcrystalline polypropylene
Color	White	White
Polymer	Polypropylene	Polypropylene
Type	Isotactic	Isotactic
Form	60- to 100-mesh powder	$<1\,\mu$m in aggregated form
Moisture, %	<0.10	<0.10
Copper, ppm	None	<50
BET surface area, m^2/g	5.4	38.0
Density, g/cm^3	0.890–0.894	0.922
X-ray crystallinity, %	45–48	70–75
Intrinsic viscosity, $[\eta]$	2.10–2.60	0.21–0.23
Molecular weight ($\overline{M}_w$)	433,700	34,800
Polydispersity index ($\overline{M}_w/\overline{M}_n$)	2.92	1.66
DSC melting endotherms (2d melt), °C	161	151, 159
Infrared spectra, ΔI (975–995 cm^{-1})	0.030–0.040	0.015–0.020

Morphological Studies of Isotactic Polypropylene and Microcrystalline Polypropylene Produced Therefrom

Only a limited amount of electron microscopy has been performed on microcrystalline polypropylenes. Several significant electron micrographs have, however, been obtained.

Figure 8.5 illustrates microcrystals recovered from an isotactic commercial polypropylene fiber, and they exhibit a gross fibrous-type morphology. Figure 8.6 shows random and clustered aggregates of unit spherical-type particles (microcrystals) recovered from a polypropylene resin precursor.

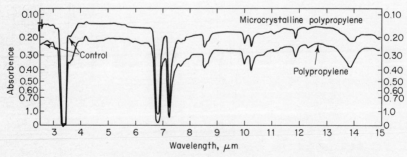

fig. 8.4   Infrared spectra of microcrystalline polypropylene and polypropylene (25).

fig. 8.5 *Microcrystals from isotactic, oriented polypropylene fibers.*

Figure 8.7 shows similar unit microcrystal particles obtained from micro-crystalline polypropylene produced from a precursor resin.   The attrition was produced by grinding the wet powder between two microscope slides vigorously for 10 min, moving the slides back and forth over the particles with pressure being supplied by the fingers and thumbs of both hands.   An aliquot

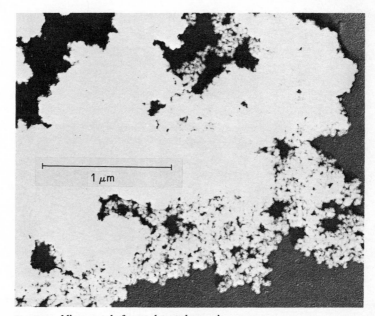

fig. 8.6 *Microcrystals from polypropylene resin.*

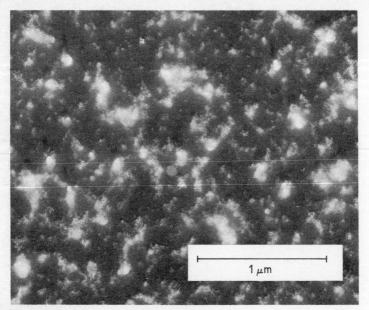

fig. 8.7 *Microcrystals of isotactic polypropylene from a resin precursor (60 min at 5 percent solids at high speed in an Osterizer, supernatant particles diluted to 0.1 percent).*

of the attrited powder was then diluted with severe attrition in a Waring Blendor, and a sample was taken for electron microscopy from the supernatant after it had stood for 16 h.

Figures 8.8 and 8.9 illustrate how the light microscope may be used to check out the degree to which the precursor polypropylene fiber has been pretreated so that the breakdown to submicron particles during the microscope-slide attrition may be followed qualitatively at least.

APPLICATIONS

Microcrystalline polypropylenes, unlike their polymer precursor, form *eutectic mixtures* with paraffin wax. This *co-crystallizing behavior* may be useful to inhibit wax migration in polymer-wax melts. Wax migration results in loss of moisture-barrier properties.

Differential calorimetric scanning curves are shown in Figs. 8.10 and 8.11. Despite the loading with paraffin wax, it is interesting to observe that the double endotherms are retained for the microcrystalline polypropylene mixtures.

The incorporation of microcrystalline polypropylene in rigid polyvinyl chloride (PVC) to improve the heat-distortion properties has been studied.

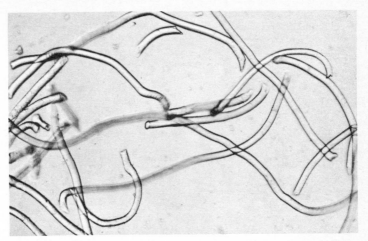

fig. 8.8 *Microcrystalline polypropylene fibers prior to attrition. (Courtesy of M. M. Cruz, Jr.)*

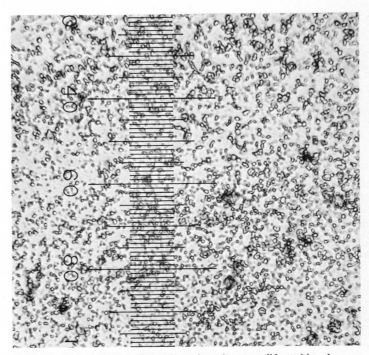

fig. 8.9 *Microcrystalline polypropylene after microscope-slide attrition, 1 division = 1.7 μm. (Courtesy of M. M. Cruz, Jr.)*

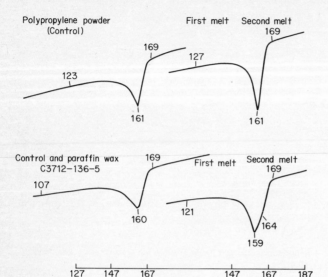

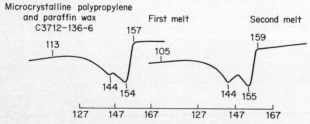

fig. 8.10 *Differential calorimetric scan of polypropylene–paraffin wax mixture* (25).

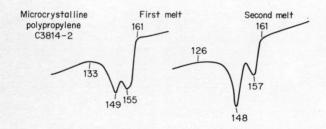

fig. 8.11 *Differential calorimetric scan of microcrystalline polypropylene–paraffin wax mixture* (25).

Inasmuch as the product has not emerged from the laboratory state of development, most uses remain to be explored. Many of the projected uses parallel those for powders or suspensoids of microcrystalline nylons or microcrystalline polyesters—with a full recognition of the lower melting point of the microcrystalline polypropylenes and their largely hydrocarbon character.

Literature Cited

1. Staudinger, H., H. Johner, R. Signer, G. Mie, and J. Hengstenberg: *Z. Phys. Chem.,* **26:**425 (1927).
2. Schlesinger, W., and H. M. Leeper: *J. Polym. Sci.,* **11:**203 (1953).
3. Battista, O. A., and P. A. Smith: Microcrystalline Cellulose, *Ind. Eng. Chem.,* **54**(9): 20–29 (1962); Colloidal Macromolecular Phenomena, *Am. Sci.,* **53**(2): 151–173 (June 1965).
4. Battista, O. A., and P. A. Smith: Level-off D. P. Cellulose Products, U. S. Patent 2,978,446 (to American Viscose Corporation), April 4, 1961.
5. Battista, O. A., et al.: Level-off Degree of Polymerization; Relation to Polyphase Structure of Cellulose Fibers, *Ind. Eng. Chem.,* **48:**333–335 (1956); "Fundamentals of High Polymers," pp. 97–113, Reinhold, New York, 1958.
6. Ranby, B. G.: Aqueous Colloidal Solutions of Cellulose Micelles, *Acta Chem. Scand.,* **3:**649–650 (1949); The Colloidal Properties of Cellulose Micelles, *Discuss. Faraday Soc.,* (11): 158–164, discussion pp. 208–213 (1951).
7. Marchessault, R. H., F. F. Morehead, and M. J. Koch: Some Hydrodynamic Properties of Neutral Suspensions of Cellulose Crystallites as Related to Size and Shape, *J. Colloid Sci.,* **16:**327–344 (1961).
8. Mukherjee, S. M., and H. J. Woods: X-ray and Electron Microscope Studies of the Degradation of Cellulose by Sulfuric Acid, *Biochem. Biophys. Acta,* **10:**499–511 (1953).
9. Battista, O. A., et al.: Colloidal Macromolecular Phenomena, Part II. Novel Microcrystals of Polymers, *J. Appl. Polymer Sci.,* **11:**481–498 (1967).
10. Wunderlich, Bernhard: *Chem.,* **39:**8–13 (October 1966).
11. Girard, A.: *Compt. rend.,* **81:**1105 (1875); *Compt. rend.,* **88:**1322 (1879).
12. Nageli, C., and S. Schwendener: *Das Mikroskop.,* 2d ed., Engelmann, Leipzig, 1877.
13. Herzog, R. O.: Cellulose Fibers, *Ber.,* **58B:**1254–1262 (1925).

14. Meyer, K. H.: *Z. Angew. Chem.,* **41:**935 (1928).

15. Meyer, K. H., and L. Misch: The Structure of the Crystalline Part of Cellulose, *V. Ber.,* **70B:**266–274 (1937).

16. Meyer, K. H., and L. Misch: The Constitution of the Crystalline Part of Cellulose, VI. The Positions of the Atoms in the New Space Model of Cellulose, *Helv. Chim. Acta,* **20:**232–244 (1937).

17. Pauling, L.: "The Nature of the Chemical Bond and the Structure of Molecules and Crystals," 3d ed., Cornell, Ithaca, N.Y., 1960.

18. Hermans, P. H., J. de Booys, and C. J. Maan: Form and Mobility of Cellulose Molecules, *Kolloid-Z.,* **102:**169–180 (1943).

19. Van der Wyk, A. J. A., and K. H. Meyer: Atomic Configuration of Cellulose, *J. Polymer Sci.,* **2:**583–592 (1947).

20. Hess, K.: Morphology and Chemistry of Organic High-Molecular Natural Substances, *Naturwiss.,* **22:**469–476 (1934).

21. Meyer, K. H., and N. P. Badenhuizen, Jr.: Transformation of Hydrated Cellulose into Native Cellulose, *Nat.,* **140:**281–282 (1937).

22. Morehead, F. F.: Ultrasonic Disintegration of Cellulose Fibers before and after Hydrolysis, *Text. Res. J.,* **20:**549–553 (1950).

23. Hermans, J., Jr.: Flow of Gels of Cellulose Microcrystals, pt. I, Random and Liquid Crystalline Gels, *J. Polym. Sci.,* pt. C, (2): 129–144 (1963).

24. Edelson, M. R., and J. Hermans, Jr.: Flow of Gels of Cellulose Microcrystals, pt. II, Effect of Added Electrolyte, *J. Polym. Sci.,* pt. C, (2): 145–152 (1963).

25. Battista, O. A., M. M. Cruz, and C. F. Ferraro: Colloidal Microcrystal Polymer Science, in Egon Matijevic (ed.), "Surface and Colloid Science," vol 3. Wiley, New York, 1971.

26. Kratohvil, S., G. E. Janauer, and E. Màtijevic: *J. Colloid Sur. Sci.,* **29**(2): 287 (1969).

27. Jayme, G., and N. Knolle: *Makromol. Chem.,* **82:**190 (1965).

28. Durand, H. W., E. G. Fleck, Jr., and G. E. Raynor, Jr.: Microcrystalline Cellulose Compositions Co-Dried with Hydrocolloids, U. S. Patent 3,573,058, March 30, 1971.

29. Tusing, T. W., O. E. Paynter, O. A. Battista: *J. Agric. Food Chem.,* **12:**284 (1964).

30. Baker, Eugene M., III: Private communication and report.

31. Richman, M. D., C. D. Fox, and R. F. Shangraw: *J. Pharm. Sci.,* **54:**447 (1965); G. E. Reier, and R. F. Shangraw: *J. Pharm. Sci.,* **55:**510 (1966).

32. Battista, O. A.: Coatings of Microcrystalline Cellulose (CCA), U. S. Patent 3,157,518, November 17, 1964.

33. Battista, O. A., and J. J. Byrne: Cellulose Crystallite Aggregates in Chromatographical Adsorption, U. S. Patent 3,179,587, April 20, 1965.

34. Battista, O. A.: Polymer Surfaces Having a Coating of Microcrystalline Cellulose (CCA), U. S. Patent 3,259,537, July 5, 1966.

35. Battista, O. A., and C. J. Boone: Chromatographic Separations by Means of Microcrystalline Cellulose (CCA), U. S. Patent 3,562,289, February 9, 1971.

36. Avicel Bulletin: Reference List, FMC Corporation.

37. Wolfrom, M. L., D. L. Patin, and R. M. de Lederkremer: *Chem. Ind.,* London, 1065 (1964).

38. Wolfrom, M. L., D. L. Patin, and R. M. de Lederkremer: *J. Chromatogr.,* **17:**488 (1965).

39. Wolfrom, M. L., et al.: *J. Chromatogr.* **18:**42 (1965).

40. Avicel Bulletin: Paints, Industrial Uses, FMC Corporation.

41. Bondi, H. S.: Glazed Ceramic Articles, U. S. Patent 3,211,575, October 12, 1965.

42. Battista, O. A.: Shaped Articles Containing Microcrystalline Cellulose (CCA), U. S. Patent 3,275,580, September 27, 1966.

43. Battista, O. A., and P. A. Smith: Formed Products of Microcrystalline Cellulose (CCA), U. S. Patent 3,278,519, October 11, 1966.

44. Battista, O. A., and J. A. Robertson: Method of Preparing Carbonized Shaped Products of Microcrystalline Cellulose (CCA), U. S. Patent 3,400,181, September 8, 1968.

45. Battista, O. A.: Carbonization of Compressed Microcrystalline Cellulose (CCA), U. S. Patent 3,639,266, February 1, 1972.
46. Veis, Arthur, J. Anesey, and S. Mussell: *Nat.,* **215:**931–934 (August 26, 1967).
47. Battista, O. A., M. M. Cruz, Jr., and W. J. Riley: Prosthetic Structures from Microcrystalline Collagen U. S. Patent 3,443,261, May 13, 1969.
48. Battista, O. A.: Method of Producing Absorbent Mats: Burn and Wound Dressings, U. S. Patent 3,471,598, October 7, 1969.
49. Battista, O. A.: Microcrystalline Collagen—An Ionizable Partial Salt of Collagen, U. S. Patent 3,628,974, December 21, 1971.
50. Battista, O. A.: Food Compositions Containing Microcrystalline Collagen, U. S. Patent 3,632,350, January 4, 1972.
51. Battista, O. A.: Water-Insoluble Microcrystalline Collagen Absorbent Mat, U. S. Patent 3,632,361, January 4, 1972.
52. Battista, O. A.: Shaped Structures Having a Self-Adherent Coating of a Water-Insoluble Ionizable Salt of Collagen, U. S. Patent 3,649,347, March 14, 1972.
53. Battista, O. A.: Microcrystalline Collagen—Foods, Pharmaceuticals, and Cosmetics Containing Same, U. S. Patent 3,691,281, September 12, 1972.
54. Battista, O. A., M. M. Cruz, Jr., and M. R. Hait: Fibrous Collagen Derived Product Having Hemostatic and Wound Binding Properties, U. S. Patent 3,742,955, July 3, 1973.
55. Yannas, I. V.: Reviews of Macromolecular Chemistry, *J. Macromol. Sci.,* **C7**(1): 49–104 (1972).
56. Hall, C. E.: *Proc. Natl. Acad. Sci. U. S.,* **42**(11): 801–806 (1956).
57. Hait, M. R., O. A. Battista, R. B. Stark, and C. W. McCord: Microcrystalline Collagen as a Biologic Dressing, Vascular Prosthesis, and Hemostatic Agent, chap. II, Wound Healing, Burns, and Infections, *Surgical Forum,* vol. XX, 1969.
58. Hait, M. R.: Microcrystalline Collagen, *Am. J. Surg.,* 330 (September 1970).
59. Lorenzetti, O. J., B. Fortenberry, E. Busby, and R. Uberman: Influence of Microcrystalline Collagen on Wound Healing, *Proc. Soc. Exp. Biol. Med.,* **140**(3): 896–900 (July 1972).
60. Rucker, I., R. Kettrey, and L. D. Zeleznick: Microcrystalline Collagen (Avitene) Film as a Substrate for Outgrowth of Primary Rabbit Kidney Fibroblasts, *Proc. Soc. Exp. Biol. Med.,* **139**(3): 749–752, (March 1972).
61. Hait, M. R., C. A. Robb, C. R. Baxter, A. R. Borgmann, and L. O. Tippett: Comparative Evaluation of Avitene Microcrystalline Collagen Hemostat in Experimental Animal Wounds, *Am. J. Surg.,* 284–287 (March 1973).
62. Wilkinson, T. S., J. H. Tenery, and D. Zufi: The Skin Graft Donor Site as a Model for Evaluation of Hemostatic Agents, *Plast. Reconstr. Surg.,* **51**(5): 541–544 (May 1973).
63. McDonald, T. O., K. Kasten, B. Britton, D. Smith, D. Bogle, A. R. Borgmann, and C. A. Robb: Biocompatibility and Bioabsorption of Avitene (Microcrystalline Collagen) Hemostat in Experimental Animals, personal communication, 1973.
64. Lorenzetti, O. J., B. Fortenberry, and E. Busby: Influence of Microcrystalline Collagen in Wound Healing II; Comparison of Several Collagen Dressings on Excised Wounds of Pigs and Rabbits, *Res. Commun. Chem. Pathol. Pharmacol.,* **5**(2): 431–440 (March 1973).
65. Erdi, N. Z., C. F. Ferraro, and O. A. Battista: Gelled Water Miscible Organic Solvents, U. S. Patent 3,393,080, July 16, 1968.
66. Battista, O. A.: Colloidal Polymer Microcrystals—New Compositions for Cosmetics, *J. Soc. Cosmet. Chem.,* **22:**561–569 (August 18, 1971).
67. Battista, O. A., and M. M. Cruz, Jr.: Light Polarizing Media, U. S. Patent 3,560,075, February 2, 1971.
68. Battista, O. A.: Photographic Coatings Containing a Water-Insoluble Ionizable Partial Salt of Collagen, U. S. Patent 3,748,142, March 14, 1972.
69. Battista, O. A.: Modified Colloidal Chrysotile, U. S. Patent 3,358,393, July 29, 1969.
70. Ferraro, C. F.: *Soc. Plast. Eng. J.,* **24**(4): 74, 99 (1968).
71. Weltmann, J.: *J. Appl. Phys.,* **14:**343 (1943).

72. Strunk, W. G.: Rheology Control Agent for Polyester Resins, *Mod. Plas.*, 166–172 (October 1967).

73. Battista, O. A.: Opening Up Asbestos with DMSO, U. S. Patent 3,410,751, November 12, 1968.

74. Battista, O. A., and F. J. Karasinski: Metallic Articles Containing Mineral Fibers, U. S. Patent 3,475,168, October 28, 1969.

75. Battista, O. A., and F. J. Karasinski: Low Melting Metallic Articles Reinforced with Modified Chrysotile, U. S. Patent 3,431,090, March 4, 1969.

76. Battista, O. A.: Stable Amylose Dispersions, U. S. Patent 3,351,489, November 7, 1967.

77. Erdi, N. Z., M. M. Cruz, and O. A. Battista: *J. Colloid Interface Sci.*, **28**(1): 36–47 (September 1968).

78. Battista, O. A.: Finely Divided Material from Synthetic Linear Polyamides, U. S. Patent 3,299,011, January 17, 1967.

79. Cranstoun, D. R., and M. M. Cruz: Method of fractionating microcrystalline colloidal polyamide suspensions, paper presented at the 156th Meeting of the American Chemical Society, Atlantic City, N.J., September 1968.

Appendix

Careful examination of all the electron micrographs presented in this treatise and in particular those delineating individual microcrystals from linear polymer precursors (i.e., cellulose, amylose, collagen, polyamide, polyester, and polypropylene) has brought into focus for the author a common denominator that may have an important relationship to the fundamental structure and morphology of crystalline linear polymers. The purpose of this epilogue is to put the results of this examination into juxtapositions in the belief that they will thereby lend themselves to a more definitive interpretation.

In essence what emerges from our analyses is that when the mechanical energy introduced to disperse microcrystals recovered from the aforementioned polymers is intense enough, the resulting smallest measurable subunits or microcrystal particles all are uniformly of about the same size and shape, with their diameters ranging from about 100 to 300 Å. This common size denominator for the smallest microcrystal entities in linear high-molecular-weight polymers seems more than coincidental. This author respectfully suggests that the most stable building block or "brick" out of which the gross architectural morphology of high-molecular-weight natural and synthetic linear polymers is fabricated has a dimension in the 100- to 300-Å size range, a truly *limiting* microcrystal size particle.

fig. A.1 *Microcrystals of polyethylene terephthalate from Dacron fibers.*

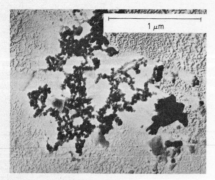

fig. A.2 *Microcrystals of polyethylene terephthalate from resin.*

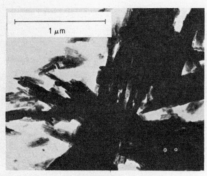

fig. A.3 *Microcrystals recovered from isotactic polypropylene fibers (normal attrition).*

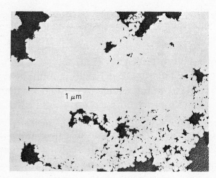

fig. A.4 *Microcrystals recovered from isotactic polypropylene resin (severe attrition).*

fig. A.5 *Microcrystals recovered from amylose after 30 min in a Waring Blendor at 10 percent solids.*

fig. A.6 *Microcrystals recovered from amylose after 60 min in an Osterizer at 10 percent solids.*

fig. A.7 *Microcrystals recovered from nylon-6 fibers.*

fig. A.8 *Microcrystals recovered from nylon-6 resin.*

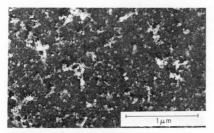

fig. A.9 *Microcrystals recovered from collagen (unheated) after 60 min in an Osterizer at 5 percent solids, temperature 25° C.*

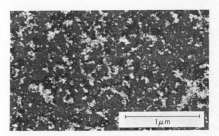

fig. A.10 *Microcrystals recovered from collagen (heated at 120° C, 30 h) after 60 min in an Osterizer at 5 percent solids, temperature 25° C.*

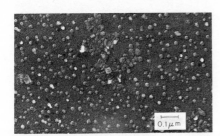

fig. A.11 *Microcrystals recovered from natural silk fibers.*

fig. A.12 *Microcrystals recovered from wool fibers.*

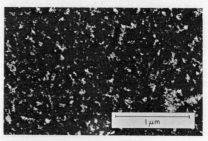

fig. A.13 *Microcrystals recovered from wood pulp after 20 min in a Waring Blendor at 5 percent solids.*

fig. A.14 *Microcrystals recovered from wood pulp after 100 min in an Osterizer at 10 percent solids.*

fig. A.15 *Microcrystals recovered from rayon after 20 min in a Waring Blendor at 5 percent solids.*

Absorbable sponges, 100
Absorbable sutures, 100
Absorbable swabs, 100
American Cyanamid Laminac 4123, 130
American Viscose Division, FMC
    Corporation, 2
Amylomaize, x-ray diffraction pattern of,
    143
Amylopectin, structure of, 139
Amylose microcrystals, 142, 143
    normal attrition, 142
    severe attrition, 142
    (*See also* Microcrystalline amylose)
Amyloses, 3
    microcrystalline (*see* Microcrystalline
        amylose)
    structure of, 138, 139
Asahi Chemical Co., 56
Asbestos, 3, 118
    chemical compositions of, 121
Asbestos fibril, morphology of, 132
Aviamides, 3
Aviamyloses, 3
Avibest SF microcrystalline silicate, 125
Avibests, 3
Avibone, 107
Avicarbon magnetic spheres, 39
Avicarbon spheres, 39

Avicarbon structures, 39
Avicels, 3
Avicon, Inc., 73
Aviesters, 3
Aviolefins, 3
Avitene synthetic bone, 107
Avitenes, 3, 113
Avory microcrystalline cellulose, 34, 36, 37,
    56
    resistance to nails, 36

Badenhuizen, N. P., Jr., 18
Battista, O. A., 20, 21, 55, 56, 97
Bauer refiner, 132
Baxter, C. R., 97
Blood-clotting agent, 100
Blood-vessel substitutes, 100
Bogle, D., 97
Borgmann, A. R., 97
Bovine-based collagen products, amino acid
    profile of, 73
Bovine cancellous bone, 102
    (*See also* Avibone)
Bovine collagen fibrils, 77
    "banana peel" effect, 77, 80, 81
    excessive swelling, 81, 82
    fat globules, 82

Bovine collagen fibrils (*Cont.*):
  wrap-around effect, 78, 79
Bovine corium collagen, composition of, 70
Bovine cortical bone, 102
Bovine tendon collagen, 59
Britton, B., 97
Brush-heap structure, 10
Bulk densities of collagen products, 75, 76
Burn and wound dressings, microcrystalline
    collagen for, 97–100
Busby, E., 97

Cab-O-Sil pyrogenic silica, 53
Calcium phosphate, 101
  mesomorphous, 103
Calcium phosphate crystals, 104
Cancellous bone, 101, 108
  bovine, 102
Caprolactam, catalytic polymerization of,
    164
Carboxymethyl cellulose (CMC), 23, 51, 52
Cellulose, 3, 12
  carboxymethyl, 23, 51, 52
  cotton, 32
  microcrystalline (*see* Microcrystalline
    cellulose)
  molecular weights by electron microscopy,
    26
  natural fibrous, in fecal specimens, 50
  viscosity average molecular weights, 26
  wood, 32
Cellulose crystals:
  peptized species of, 6
  x-ray diffraction data, 36
Chromatography:
  gel permeation, 160
  thin-layer, 55
Chromatoplates, preparation of, using
    microcrystalline cellulose, 55–56
Chrysotile, 118, 132
  water of composition, 125
Chrysotile fibrils, 119
  channels or cores, 134
  decomposition by electron beam, 134
Coatings, collagen, 113–114
Collagen, 3, 74
  amino acids, 59
  colloidal fragments, 60
  definition of, 58–59

Collagen (*Cont.*):
  imino acids, 59
  IR (infrared) spectra, 59
  "limiting microfibrils," 60
  microcrystalline (*see* Microcrystalline
    collagens)
  periodicity, 59
  specific optical rotation, 59
Collagen coatings, 113–114
Collagen fibers:
  bovine (*see* Bovine collagen fibrils)
  tensile strength of, 59
Collagen gels, role of pH on properties of,
    92
Collagen products:
  bovine-based, 73
  bulk densities of, 75, 76
Collagen sutures, natural, 70
Conventional polymer technology, schematic
    diagram, 24
Corium collagen:
  bovine, composition of, 70
  ethyl alcohol–water extractions, 61
Corium collagen fibers, 77
Cortical bone, 101, 107
  bovine, 102
Cosmetics, use of microcrystalline collagens
    in, 111–112
Cotton cellulose, level-off DP, 32
Cotton microcrystals, 33
Cowles dissolver, 129
Cowles Hi-Shear Mixer, 123
Cranstoun, D. R., 153
Cruz, M. M., Jr., 153, 191

de Booys, J., 18
Denaturation, 200 proof alcohol, 77
Differential calorimetric scans, 154–158
  of microcrystalline polyamides, 156–158
  of microcrystalline polypropylene, 187
  of polypropylene powder, 187
Dimethyl sulfoxide, 106–109
Dioctyl phthalate (DOP), 128
Dispersions, microcrystalline collagen,
    106–111
Dissolver, Cowles, 129
DLVO (Derajaguin-Landau-Verwey-
    Overbeek) theory, 19
Dow-Corning Silicon 200, 172

Edelson, M. R., 19, 41
Egenomen, 58

Ferraro, C. F., 127
Foods:
   use of microcrystalline cellulose in (*see*
      Microcrystalline cellulose, uses of)
   use of microcrystalline collagen in, 111,
      112
Fortenberry, B., 97

Gelatin:
   amino acid composition, 72
   electron microscopy, 86
   photographic, 116
Graham, Thomas, 1
Gutta-percha, 2

Haake Rotovisco Viscometer, 159
Hait, M. R., 97
Heat shock, control of, in frozen desserts, 3,
   52
Hemostat adhesive, microcrystalline
   collagen, 64, 95, 96
Hemostatic agent, 100
Hermans, J., Jr., 19, 41
Hermans, P. H., 18
Herzog, R. O., 17
High-shear mechanical attrition, 123
Hobart mixer, 38
Human tendon collagen, amino acid
   composition of, 74
Hydrocellulose, 17

Janauer, G. E., 19, 42, 43
Jayme, G., 20
Johner, H., 2

Kasten, K., 97
Kettrey, R., 97
Knolle, N., 20
Kolla, 1, 58
Kratohvil, S., 19, 42, 43

Laminac 4123, American Cyanamid, 130
Leeper, H. M., 2

Level-off degree of polymerization (LODP)
   cellulose from rayon, 5
Limiting microcrystal sizes, 197–200
Liquid crystalline phase of microcrystalline
   particles, 41
LODP (level-off degree of polymerization)
   cellulose from rayon, 5
Lorenzetti, O. J., 97

Maan, C. J., 18
McCord, C. W., 97
McDonald, T. O., 97
Magnesium silicate (asbestos), 3
Marchessault, R. H., 18, 19
Matijević, Egon, 19, 42, 43
Mechanical attrition, high-shear, 123
Mesomorphous calcium phosphate, 103
Meyer, K. H., 17, 18
Microcrystal polymer products, 3
Microcrystal polymer technology:
   pertinent size ranges, 34
   schematic diagram, 25
Microcrystalline amylose gels, yield-stress
   properties of, 141, 145
Microcrystalline amyloses, 138–145
   applications of, 141–143
   iodine, nonabsorption, 141
   preparation of, 139–141
   properties of, 141
   raw materials for, 139
   recrystallized, x-ray diffraction pattern of,
      144
   x-ray diffraction pattern of, 144
Microcrystalline cellulose gels:
   role of counterions, 43
   role of pH on viscosity, 40
   salt effects of, 41
   thixotropy properties of, 42
   yield values of, 42
Microcrystalline celluloses, 17–57
   ablation resistance of, 37
   Avory, 34, 36, 37, 56
   buildup of free microcrystals, 39, 40
   bulk density, 23
   carbon spheres, 37
   carbonized forms, 39
   catalyst formers, 56
   in ceramic glazes, 56
   chemical and physical properties of, 24
   CMC blends, 23, 51, 52

Microcrystalline celluloses (*Cont.*):
  column chromatography, 55
  commercial applications of, 45–57
  compressed shapes of, 34
  conversion of oily or sticky products to
    free-flowing forms, 31
  from cotton linters, 22
  dried gel forms of, 34, 37
  in dripless paints, 56
  in fecal specimens, 48–50
  in foods (*see specific food under* uses of
    below)
  granulation properties of, 31
  graphite forms of, 37
  heat resistance properties of, 37
  in the human digestive process, 48–50
  industrial uses of, 54–55
  instantly dispersible blends, 23
  labelled with C^{14}, 48–50
  level-off DP, 20
  oxidized forms of, 43, 44
  polymolecularity of, 34
  preparation of, 21–23
  properties of, 23–43, 52
    carbon spheres, 38
    refractive indices, 32
  sodium carboxymethyl cellulose
    derivative, 43, 44
  solubility in NaOH, 23
  spherical catalyst substrates, 57
  spheroidized forms of, 38
  structural carbon forms of, 37
  structural forms of, 37
  thin-layer chromatography, 55
  topochemical derivatization, 43, 44
  uses of: in canned tuna salads, 45
    in ceramics, 46
    in decorative laminates, 46
    in free-flowing cheese, 48
    in free-flowing peanut butter, 48
    in frozen desserts, 45
      to control heat shock, 52
    in heat-stable cream sauces, 51
    in hollandaise sauces, 45
    in ice creams, 52
    in ice milks, 52
    in imitation processed-cheese spreads,
      51
    in low-calorie dressings, 51
    in low-calorie foods, 46
    major, 3–4

Microcrystalline celluloses, uses of (*Cont.*):
    in no-calorie imitation butter, 47
    in nondairy coffee creamers, 51
    in pharmaceutical tablet products,
      52–54
    in pharmaceuticals, 45, 46, 52–54
    in rayon tire yarns, 46, 47
    in shelf-stable foods, 51
    in sour cream, 51
    as a tablet excipient, 46
    in toppings, 51
    in water-base paints, 46
  from viscose rayon, 21
  water absorption, 23
  from wood-pulp-alpha cellulose, 23, 24
Microcrystalline collagen bone, preparation
    of: in cartilage-like form, 105
  in cortical form, 105
Microcrystalline collagen films, mechanical
    properties of, 110
Microcrystalline collagen flour, 64, 69, 89,
    90
Microcrystalline collagen gels, 76
  effect of absorbed HCl, 93
  freeze-dried aqueous, 97, 98
  organic solvent systems, 106, 108, 109
Microcrystalline collagen mats, freeze-dried,
    99–100
Microcrystalline collagens, 58–117
  as an anchoring agent, 114
  apatite synthetic bone structures, 101–106
  applications of, 95–117
  Avibone, 107
  as a bioengineerable material, 100
  bound HCl, 63
  cancellous-type bone, 108
  in coatings, 113, 114
  color changes with heat and cross-linking
    agents, 100
  in cosmetics, 111
  and cross-linking agents, 99
  definition of, 64
  demonstrated and projected uses of, 117
  dispersions, colloidal, 106–111
  as an egg-white substitute, 112
  as a film-coating material, 114
  in flour form, 64, 69, 89, 90
  in foods, 111, 112
  in freeze-dried forms, 69
  freeze-dried gels: aqueous, 97, 98
    aqueous-alcohol, 98

Microcrystalline collagens (*Cont.*):
in fused-fibrous form, 90
gel, 1 percent aqueous, 93
gel viscosities in water, 76
and glycerine, 76
gels (water-organic solvents), 97
heat sterilization, 88–89
in hemostat-adhesive form, 64, 95, 96
in hemostat form soaked with blood, 96
importance of initial compression for
hemostasis, 96
IR spectrum: after deuteration, 94
before deuteration, 94
in lint-free products, 99
major uses of, 4
medical grade, 71
medical uses (chart), 100
microbiology, 71
microcrystals: fibril precursor, 86
after heat sterilization, 88, 90
normal attrition, 87
overattrition, 87, 89, 91
in nonhemostatic fibrous form, 97
in nonwoven web form, 95
optimum HCl concentrations, 93
partial acid salts (other than HCl), 67
pH versus viscosity, 68
in pharmaceuticals, 111
in photographic emulsions, 113–115
preparation of, 61–69
properties of, 69–90
after heat treatment, 76
before heat treatment, 76
role of alum in, 100
in sheet (nonwoven) form, 95
shrinkage with heat, 99
spherical microcrystals, 83, 88
surface areas, 76
surgical uses of, 95, 96, 99
suspensoids, preparation of, 66
synthetic bone, 104
synthetic bone structures, 107, 108
apatite, 101–106
water-glycerine gels, 113
Microcrystalline nylons:
major uses of, 4, 164
properties of, 153
Microcrystalline polyamide gels:
dependence of properties on free
microcrystals, 162
effect of freeze-thaw cycle, 161

Microcrystalline polyamide gels (*Cont.*):
fractionation procedure, 154
rheological properties of, 159
surface tension, 161
temperature insensitivity, 161
viscosity-pH dependence, 159, 160
yield-stress properties of, 161
Microcrystalline polyamides, 149–168
in aluminum coatings, 168
as anchoring agents, 166
applications of, 164–168
in aqueous gel form, 153
ash in, percent of, 154
bulk density of, 154
differential scanning data, 154–158
for dilution of high-solids pastes, 153
electrical conductance, 164, 166, 167
electron diffraction pattern of, 168
electrophoretic mobility of, 166
flowsheet of process, 152
gel permeation chromatography, 160
glass-cloth adhesives, 167
glass-cloth fabrics, 164
glass fibers, 164
intrinsic viscosity of, 154
melting point of, 154
membranes, 168
microcrystal deaggregation of, 155
moisture regain of, 154
molecular weight of, 154, 157–159
nitrogen in, percent of, 154
in paints, 167
platelet microcrystals, 164
polydispersity index, 154
preparation of, 150
properties of, 152, 154
as sizing agents, 166, 167
stable suspensions, 164
surface area (BET), 154
ultrathin films, 168
viscosity versus shear, 165
x-ray diffraction pattern of, 165
Microcrystalline polycarbonate, 174
microcrystals, 175
preparation of, 174
Microcrystalline polyester gels:
role of pH, 177
viscosities, 176, 177
zeta potentials, 178
Microcrystalline polyesters, 169–181
acid resistance of, 180

Microcrystalline polyesters (*Cont.*):
  alkali resistance of, 180
  applications of, 180–181
  aqueous gels, 173
  in coatings, 181
  commercial processes for, 169
  in films, 181
  fluidized-bed powders, 181
  Grade R (regular grade), 176
  Grade SF (superfine grade), 176
  major uses of, 4–5
  microcrystals, 173
    morphology of, 178
    from overattrited gels, 180
  from PET fibers, 169, 173–176, 178, 180
  physical and chemical properties of, 179
  preparation of, 170–172, 175
  properties of, 175–180
  raw material annealing, 172
  shaped articles, 180
  x-ray diffraction pattern of, 175
Microcrystalline polyolefins, 182–192
  differential calorimetric scans of, 187
  major uses of, 5
  molecular weight of, 184
  molecular-weight distribution, 185
  preparation of, 183–184
  properties of, 184–190
  refractive index, 186
  x-ray diffraction pattern of, 185
Microcrystalline polypropylene:
  applications of, 190–192
  differential calorimetric scans with
      paraffins, 192
  eutectic mixtures with paraffin wax,
      190
  gel preparation from, 184
  infrared spectrum, 188
  microcrystals: from fibers, 189
    from overattrited gels, 190
    from resins, 189
  stable colloidal suspensoids of,
      preparation of, 183
  table of properties, 188
Microcrystalline polypropylene fibers, 191
  after attrition, 191
Microcrystalline silicate fibrils, SiO_2
    coatings, 133
Microcrystalline silicate gels:
  diethylene glycol synergism, 131
  pseudoplastic properties of, 129

Microcrystalline silicate gels (*Cont.*):
  shear-stress rate of, 127
  viscosity change with time, 131
Microcrystalline silicates, 118–137
  applications of, 133–137
  aqueous suspensoid, 120
  Avibest SF, 125
  ball-milled microcrystals, 124
  chemical composition of, 124–126
  as a cigarette filter, 136
  comparison with pyrogenic silica, 135
  composition of, 125
  decolorization of, 135
  and dimethyl sulfoxide, 132
  in DOP (dioctyl phthalate), 128
  foamed aluminum, 136
  graphite greases, 136
  magnetic clutch greases, 136
  magnetite greases, 136
  major uses of, 4
  microcrystals, 123
    ball-milled, 124
  modified chrysotile surfaces, 121
  physical properties of, 124–126
  in pigmented paints, 126
  in plastisols, 128
  in polyester resins, 129, 130
  in polyhydroxylic liquids, 129
  potential uses of, 137
  preparation of, 120–124
  for reinforcement of soft metals, 137
  rheology of, in organic liquids, 127–128
  as a rheology control agent, 129
  surfactants, 119
  thickening action of, 126
  in tricresylphosphate, 127
  viscosities of ethylene glycol gels, 122
  for water purification, 135
  x-ray diffraction pattern of, 125
Microcrystalline starches, major uses of, 4
Microcrystals:
  from amylose starch: regular attrition, 198
    severe attrition, 198
  of cellulose-dislodging, 27
  from collagen: after heat sterilization and
      severe attrition, 199
    severe attrited, unheated, 199
  from natural silk, 199
  from nylon-6 fibers, 199
  from nylon-6 resin (pellets), 163, 164, 199
  from polyester fibers, 198

Microcrystals (*Cont.*):
  from polyester resin, 198
  from polypropylene fibers, 198
  from polypropylene resin, 198
  from viscose rayon, 200
  from wood pulp: normal attrition, 200
    severe attrition, 200
  from wool, 199
Misch, L., 17
Morehead, F. F., 18
Mukherjee, S. M., 18

Nageli, C., 17
Natural collagen sutures, composition of,
  70
Natural fibrous cellulose in fecal specimens,
  50
Natural silk, 3, 199
  microcrystals from, 84
Nylon-6 platelet microcrystals, 164
Nylon tire yarns, 46–47
Nylon-12 powder particles, 163
Nylons (*see* Microcrystalline nylons)

Orthopedic surgery, synthetic-bone
  materials and, 106

Particle-network model, 41
Particle sizes of various forms of celluloses,
  34
Pauling, L., 18
Peptized species of cellulose crystals, 6
Perkin-Elmer DSC-1 differential scanning
  calorimeter, 154
Pharmaceuticals:
  use of microcrystalline cellulose in, 45, 46,
    53
  use of microcrystalline collagen in, 111
Photographic emulsions, 113–116
  composition of, 116
Polyamides (*see* Microcrystalline
  polyamides)
Polycarbonate (*see* Microcrystalline
  polycarbonate)
Polycarbonate monomer, 170
Polyesters (*see* Microcrystalline polyesters)
Polyethylene terephthalate (PET), 169,
  173–176, 178, 180

Polyolefins (*see* Microcrystalline polyolefins)
Polypropylene:
  infrared spectrum of, 188
  microcrystalline (*see* Microcrystalline
    polypropylene)
Polyvinyl chlorides, 12
Prerequisites, microcrystal polymers, 8
Prostheses, 106
Pseudoplastic-rheopectic behavior, 10
Pseudoplasticity, 41
Pulverized wood pulp, 5
Pyrogenic silica, 126, 130

Ranby, B. G., 18
Rayon fibrils, ultrafine, 27
Rayon microcrystals, 32
Readco Extructor, 132, 153
Recrystallized amylose lamellae
  microcrystals, 143
Refiner, Bauer, 132
Rietz Extructor, 87–89, 123
Robb, C. A., 97
Rucker, I., 97

Schlesinger, W., 2
Signer, R., 2
Silica spheres, coating on microcrystalline
  silicate fibrils, 133
Silicates (*see* Microcrystalline silicates)
Silk, natural, 3, 199
  microcrystals from, 84
Smith, P. A., 20, 21, 47, 97
Starches, microcrystalline, major uses
  of, 4
Stark, R. B., 97
Staudinger, H., 2
Strong-Cobb hardness data, 54
Strunk, W. G., 129
Synthetic bone, 100–106
  Avitene, 107
  chemical composition of, 101
  cortical form of, 106, 107
  cross-linking agent, 109

Tenery, J. H., 97
Tropocollagen, 60, 85
Tropocollagen molecules, 83

Uberman, R., 97
Upjohn Co., 73
USP disintegration apparatus, 54

Van der Wyk, A. J. A., 18
Veis, A., 60
Viscose seeding experiment, 47

Weltmann, J., 127
Wilkinson, T. S., 97
Wolfrom, M. L., 55
Wood pulp, pulverized, 5

Wood-pulp-cellulose microcrystals:
    overattrition, 85
    regular attrition, 33
Wood-pulp fibrils, 26
Woods, H. J., 18
Wool, microcrystals of, 84
Wound dressings, microcrystalline collagen
    for, 97–100
Wunderlich, B., 11

Zeleznick, L. D., 97
Zufi, D., 97